Basic Stereochemistry of Organic Molecules

Basic Stereochemistry of Organic Molecules

Rose Marie O. Mendoza

ARCLER
PRESS

www.arclerpress.com

Basic Stereochemistry of Organic Molecules

Rose Marie O. Mendoza

Arcler Press
224 Shoreacres Road
Burlington, ON L7L 2H2
Canada
www.arclerpress.com
Email: orders@arclereducation.com

ISBN: 978-1-77407-792-4 (Hardcover)

ABOUT THE AUTHOR

Dr. Rose obtained her PhD in Chemical Engineering from the University of the Philippines-Diliman in 2013, Her Masters in Chemical Engineering and BS Chemical Engineering degree from Adamson University. She is also a Professor in the Graduate School Department under the Master of Engineering Program at Adamson University since 2006, and is a Visiting Research Fellow at the Department of Environmental Engineering and Science, and the Department of Environmental Resource Management in Chia Nan University of Pharmacy and Science, Tainan Taiwan since 2010. She obtained her Post Doctorate Degree in Green Power: Hydrogen Generation and Fuel Cell Development from the University of California Merced, USA (in collaboration with the Energy Storage and Conversion Materials Laboratory of the University of the Philippines, Diliman) and is presently also a collaborating researcher at the Kindai University, Nara Japan, focusing on applications of electric and electrostatic field.

Dr. Mendoza has several international and local researches covering Fuel cells and solid oxide electrolysis cells (SOECs), green energy and green technology, biomedical and natural products, fuels and fuel development, at water and wastewater treatment, recovery and remediation. She also published several international textbooks on water management, introductory chemical engineering and chemistry. Presently, Dr. Mendoza is the Chief Science Officer of Mark-Energy Revolution Corporation, Philippines.

TABLE OF CONTENTS

GLOSSARY

A

Absolute Configuration – An absolute configuration refers to the spatial arrangement of the atoms of a chiral molecular entity and its stereochemical description, e.g., R or S, referring to Rectus, or Sinister, respectively.

Atom – is the smallest constituent unit of ordinary matter that constitutes a chemical element.

Axial position – location in a trigonal bipyramidal geometry in which there is another atom at a 180° angle and the equatorial positions are at a 90° angle.

Achiral – (chiefly of a molecule) symmetric in such a way that it can be superimposed on its mirror image; not chiral.

Adaptability – the quality of being able to adjust to new conditions.

Auxiliary – providing supplementary or additional help and support.

B

Bond Angle – angle between any two covalent bonds that share a common atom.

Bond Dipole Moment – separation of charge in a bond that depends on the difference in electronegativity and the bond distance represented by partial charges or a vector.

Bond Distance – (also, bond length) distance between the nuclei of two bonded atoms.

Bond Energy – (also, bond dissociation energy) energy required to break a covalent bond in a gaseous substance.

Bond Length – distance between the nuclei of two bonded atoms at which the lowest potential energy is achieved.

C

Carbides – a carbide usually describes a compound composed of carbon and a metal.

Carbohydrates – is a biomolecule consisting of carbon, hydrogen, and oxygen atoms, usually with a hydrogen – oxygen atom ratio of 2:1.

Cellulose – is a polysaccharide consisting of a linear chain of several hundred to many thousands of β linked D – glucose units.

Chemistry – the scientific discipline involved with elements and compounds composed of atoms, molecules, and ions: their composition, structure, properties, behavior, and the changes they undergo during a reaction with other substances.

Chiral – asymmetric in such a way that the structure and its mirror image are not superimposable.

Chirality – In chemistry, a molecule or ion is called chiral if it cannot be superposed on its mirror image by any combination of rotations and translations. This geometric property is called chirality.

Configuration – an arrangement of parts or elements in a particular form, figure, or combination.

Conformational – is a form of stereoisomerism in which the isomers can be interconverted just by rotations about formally single bonds.

Correlation – Correlation is a statistical technique that can show whether and how strongly pairs of variables are related.

Covalent Bond – covalent bond, also called a molecular bond, is a chemical bond that involves the sharing of electron pairs between atoms.

Cyanides – is a chemical compound that contains the group C≡N.

Congruent – in agreement or harmony.

Composition – the nature of something's ingredients or constituents; the way in which a whole or mixture is made up.

D

Diastereomers – Diastereomers are stereoisomers that are not mirror images of one another and are non – superimposable on one another. Stereoisomers with two or more stereocenters can be diastereomers. It is sometimes difficult to determine whether or not two molecules are diastereomers.

Diastereotopic Ligands – when the replacement test is applied to two like ligands in a molecule, if the resultant molecules are diastereomers, the two ligands are said to be diastereotopic.

Disaccharides – is the sugar formed when two monosaccharides are joined by glycosidic linkage.

Deliberately – consciously and intentionally; on purpose.

E

Electron-pair Geometry – arrangement around a central atom of all regions of electron density (bonds, lone pairs, or unpaired electrons).

Enantiomer – is one of two stereoisomers that are mirror images of each other that are non-superposable, much as one's left and right hands are mirror images of each other that cannot appear identical simply by reorientation.

Enantiotropic – the stereochemical term enantiotropic refers to the relationship between two groups in a molecule which, if one or the other were replaced, would generate a chiral compound. The two possible compounds resulting from that replacement would be enantiomers.

Enzyme catalysis – it is the increase in the rate of a process by a biological molecule, an "enzyme."

Equilibrium – is the state in which both reactants and products are present in concentrations which have no further tendency to change with time, so that there is no observable change in the properties of the system.

Endowed – give or bequeath an income or property to (a person or institution).

F

Fluids – a fluid is a substance that continually deforms under an applied shear stress, or external force.

G

Galactose – it sometimes abbreviated Gal, is a monosaccharide sugar that is about as sweet as glucose, and about 65% as sweet as sucrose.

Glycogen – is a multibranched polysaccharide of glucose that serves as a form of energy storage in animals, fungi, and bacteria.

H

Handedness –is a better, faster, or more precise performance or individual preference for use of a hand, known as the dominant hand.

Heterotopic – atoms that are structurally different from one another are heterotopic to one another and are examples of constitutional isomers.

Homotopic Ligands – like ligands in an organic molecule that are equivalent under all conditions are said to be homotopic.

Hydroxyl – hydroxy or hydroxyl group is the entity with the formula OH. It contains oxygen bonded to hydrogen.

I

Inorganic compounds – typically a chemical compound that lacks carbon – hydrogen bonds, that is, a compound that is not an organic compound.

Interconvertible – convertible the one into the other.

Ionic Bond – strong electrostatic force of attraction between cations and anions in an ionic compound.

Isomer – each of two or more compounds with the same formula but a different arrangement of atoms in the molecule and different properties.

Imprecision – lack of exactness or accuracy.

Inherently – in a permanent, essential, or characteristic way.

L

Lactose – is a sugar composed of galactose and glucose subunits and has the molecular formula $C_{12}H_{22}O_{11}$.

Lewis Structure – diagram showing lone pairs and bonding pairs of electrons in a molecule or an ion.

Ligands – is an ion or molecule that binds to a central metal atom to form a coordination complex.

Lipids – a lipid is a biomolecule that is soluble in nonpolar solvents.

Lone Pair – two (a pair of) valence electrons that are not used to form a covalent bond.

M

Maltose – also known as maltobiose or malt sugar, is a disaccharide formed from two units of glucose joined with an α bond.

Molecular Geometry – Molecular geometry is the three-dimensional arrangement of the atoms that constitute a molecule.

Molecular Structure – structure that includes only the placement of the atoms in the molecule.

Molecular Symmetry – Molecular symmetry in chemistry describes the symmetry present in molecules and the classification of molecules according to their symmetry.

Monosaccharide – also called simple sugar, are the simplest form of sugar and the most basic units of carbohydrates.

N

Nucleic Acids – are the biopolymers, or small biomolecules, essential to all known forms of life.

O

Octet Rule – guideline that states main group atoms will form structures in which eight valence electrons interact with each nucleus, counting bonding electrons as interacting with both atoms connected by the bond.

Overlapping – extend over so as to cover partly.

P

Petrochemical – are the chemical products obtained from petroleum by refining.

Pharmacologists – a specialist in the science of medications.

Polarimeter – a scientific instrument used to measure the angle of rotation caused by passing polarized light through an optically active substance.

Polarized – restrict the vibrations of (a transverse wave, especially light) wholly or partially to one direction.

Polysaccharides – are long chains of carbohydrate molecules, specifically polymeric carbohydrates composed of monosaccharide units bound together by glycosidic linkages.

Prochirality – a molecule or atom is prochiral if it contains, or is bonded to, two constitutionally identical ligands (atoms or groups), replacement of one of which by a different ligand makes the molecule or atom chiral.

Proteins – Proteins are large biomolecules, or macromolecules, consisting of one or more long chains of amino acid residues.

Portrays – depict (someone or something) in a work of art or literature.

R

Racemic – composed of dextrorotatory and levorotatory forms of a compound in equal proportion.

Relative Configuration – Relative configuration refers to the configuration of a molecule in relation to other atoms on the same molecule, in relation to other molecules, or in relation to another form of the same molecule.

Reactivity – Reactivity is the tendency of a substance to undergo chemical reaction, either by itself or with other materials, and to release energy.

Receding – go or move back or further away from a previous position.

Refractive – of or involving refraction.

Retrieval – the process of getting something back from somewhere.

S

Stereochemistry – a subdiscipline of chemistry, involves the study of the relative spatial arrangement of atoms that form the structure of molecules and their manipulation.

Stereogenic center – a particular instance of a Stereogenic element that is geometrically a point.

Stereogenic Components – a Stereogenic element is a center, axis or plane that is a focus of stereoisomerism, such that an interchange of two groups attached to this feature leads to a stereoisomer. Stereogenic elements may be chiral or achiral.

Stereoselective – (of a reaction) preferentially producing a particular stereoisomeric form of the product, irrespective of the configuration of the reactant.

Substituent – is an atom or group of atoms which replaces one or more hydrogen atoms on the parent chain of a hydrocarbon, becoming a moiety of the resultant new molecule.

Sucrose – is common sugar. It is a disaccharide, a molecule composed of two monosaccharides: glucose and fructose.

Symmetry – the quality of being made up of exactly similar parts facing each other or around an axis.

Symmetry element – A symmetry element is a point of reference about which symmetry operations can take place. In particular, symmetry elements can be identities, mirror planes, axes of rotation (both proper and improper), and centers of inversion.

Symmetry Operators – In the context of molecular symmetry, a symmetry operation is a permutation of atoms such that the molecule or crystal is transformed into a state indistinguishable from the starting state.

T

Three-dimensionality – a geometric setting in which three values are required to determine the position of an element. This is the informal meaning of the term dimension.

Topicity – in stereochemistry, Topicity is the stereochemical relationship between substituents and the structure to which they are attached. Depending on the relationship, such groups can be heterotopic, homotopic, enantiotropic, or diastereotopic.

Terminology – the body of terms used with a particular technical application in a subject of study, profession, etc.

Topology – the study of geometrical properties and spatial relations unaffected by the continuous change of shape or size of figures.

Trefoil – a small European plant of the pea family, with yellow flowers and three-lobed clover-like leaves.

Triggered – (of a mechanism) activated by a trigger.

LIST OF FIGURES

LIST OF TABLES

LIST OF ABBREVIATIONS

CA	chemical abstracts
CIP	Cahn-Ingold-Prelog
DFT	density-functional theory
DNA	deoxyribonucleic acid
EDCI	1-ethyl-3-(3-dimethylaminopropyl)carbodiimide
FDA	Food and Drug Administration
GAP	glyceraldehyde-3-phosphate
He	helium
HOBT	1-hydroxybenzotriazole
IPP	isopentenyl diphosphate
IUPAC	International Union of Pure and Applied Chemistry
MIMs	mechanically interlocked molecules
MSMs	mechanically self-locked molecules
N	nitrogen
NaCl	sodium chloride
NADH	nicotinamide adenine dinucleotide
NMR	nuclear magnetic resonance
O	oxygen
ORD	optical rotatory dispersion
PIM	prostereoisomerism
RNA	ribonucleic acid
S	sulfur
TECRDB	thermodynamics of enzyme-catalyzed reactions database
THYME	tetra hydroxyl methyl ethylene
TLC	thin-layer chromatographic
USCI	unit-subdued-cycle-index
VSEPR theory	valence-shell electron-pair repulsion theory

PREFACE

Organic chemistry deals with the various kinds of hydrocarbons and their arrangements and reveals the possibilities that can be considered by reacting one hydrocarbon with other compounds, especially their reaction with oxygen-containing substances.

Organic chemistry opens up new dimensions in the field of chemistry and finds a wide range of applications across several domain. As such organic chemistry has been found very useful in medicines and pharmaceutical products including cosmetics, aromatic compounds, composites, manufacture of various materials and products, as well as fuels and fuels processing for several industries, food, and shelter and many more.

This book aims to educate the readers about the manner in which the various molecules in the organic compounds are arranged and the formations they undergo when they are subjected to certain conditions. The book basically explains the stereochemistry of the various kinds of organic molecules so that the readers can equip themselves with the knowledge on the subject matter.

The book begins by explaining the historical significance of stereochemistry to the readers, by making them aware of the role of isomers and the process of stereoisomerism. It also explains the basic applied stereochemistry in isomers.

The book then moves on to explaining the different concepts and terminologies in organic chemistry that may help the readers in getting a proper gasp of the book in the later phases and in understanding the various phenomena and technical terms associated with them.

The book further goes on to explain the various principles that guide the stereochemistry of organic molecules and the applications of those principles. It also relates the significance of the principles in the field of organic chemistry and the molecules involved in it.

Then, the book talks about the conformation of the acyclic and the cyclic molecules and explains the processes involved in such conformations. The readers are also educated of the effects that are observed on the conformations and the manner in which these conformations can be used to yield desired compounds.

The book also explains the readers about the way in which different molecules may exhibit symmetry in their appearance and arrangement and the significance of such kind of symmetry with respect to their physical characteristics and the strength of the compounds and molecules.

Following this, the book explains the processes involved in the separation of the stereoisomers and the resolution and racemization involved with the molecules. The readers are further taken into the details of the use of stereochemistry in chemical reactions.

The book concludes itself by talking about the topicity and the pro-stereoisomerism, explaining their role in the stereochemistry of organic molecules and the relationships they hold in the domain. It is of great aspiration that this book will be able to provide the knowledge and the information needed by the readers to understand stereochemistry and its importance in understanding the organic chemistry of molecules and compounds covered.

Chapter 1

Overview of Stereochemistry of Organic Molecules

CONTENTS

Stereochemistry is the branch of chemistry that explains the arrangements of atoms of the organic molecules involved in the chemical reaction. This chapter starts with the historical development of stereochemistry and sheds some light on isomers and the importance of their stereochemistry as well as their participation in the process known as stereoisomerism. There are several types of isomers such as enantiomers and diastereomers that are introduced and explained briefly in this chapter. Asymmetric syntheses and various types of stereochemistry, such as static stereochemistry and applied stereochemistry, is also explained towards the end of this chapter.

1.1. INTRODUCTION

Stereochemistry is one of the most important and vastly used area of organic chemistry. This branch of chemistry deals with the three-dimensional arrangements in space of atoms in the molecule. Sometimes, it is also referred to as 3-D chemistry since the prefix "stereo-" used in the term refers to the "three-dimensionality of a system." Stereochemistry is very important in the fields of inorganic and organic chemistry and has applications in the practice of medicine and allied fields, drugssw preparation, chemical synthesis and manufacturing.

Metal complexes are readily formed by transition metals with various ligands. For instance, the square planar complex of platinum cis-$PtCl_2(NH_3)_2$ (cisplatin) is considered as an anti-cancer agent. On the other hand, the trans-isomer is inactive against cancer (that is it is toxic in nature), and so it is not useful in chemotherapy.

Generally, nature is very selective in the formation of one set of stereoisomers of natural products. Enzymes heavily mediate this selectivity. The stereo chemistry of an organic molecule is also strongly considered with respect to the double bonds because of the limited rotation around this bond.

Chirality of molecules play an important role in stereochemistry. There are several groups of organic compounds that are mirror image of each other. Looking at the chirality and handedness, it will be easier to visualize this by using your hands. If look at your hands the left hand is the mirror image of your right hand; this means that if you put your left hand over a mirror, the image in the mirror will be the same as your right hand (*see* Figure 1.1). This means that there is almost 100% handedness between the left hand and the right hand. However, if you superimposed your left hand over your right hand (*see* Figure 1.2), they will not trace the same pattern. Chiral molecules, therefore can be compared to your left hand and right hand, they are mirror image of each other, meaning they have the same components, but they are superimposable meaning there components will not be exactly matched with each other if you try to stack one over the other; because they have different conformation or structure and they lack point of symmetry. This means that no matter at what angle you rotate one of them, they will never become superimposable.

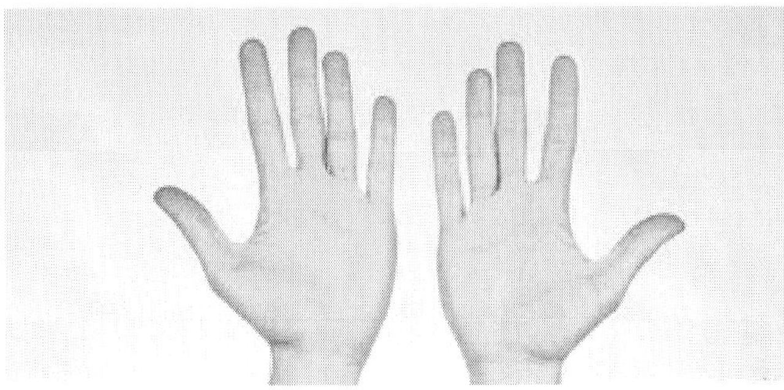

Figure 1.1. Handedness or chirality illustrated using your hands.

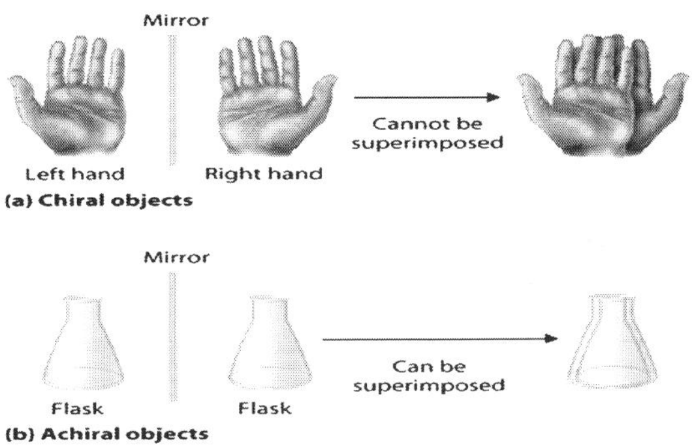

Figure 1.2. Left hand and right hand, however, is not superimposable.

There is some amount of innate chirality present in the nature. Chirality can be observed among organic compounds are molecules. To name some, the α-amino acids in proteins and enzymes, sugars in cellulose and nucleotides of DNA are chiral in nature and may seem to be in pure forms. Therefore, we can say that majority of biological molecules or biomolecules utilizes their chirality to interact and assume their biologic functions.

Isomers, on the other hand, plays an important role in recognizing chirality and understanding further that mechanism of stereoisomerism. To define first, isomers are molecules that have the same molecular formula (composition of elements), but they have different properties. If we consider the molecular make-up of the hands in Figures 1.1 and 1.2, our left and right hand are isomers of each other. Structural isomers have the same molecular formula (that is the composition of elements), but they have different structural formula (that is bond connections).

Though there is no particular requirement for structural isomers to be interconvertible, there exist some of the special cases in which the compounds are in equilibrium. One such case is Keto-enol tautomerism.

1.2. HISTORICAL STUDY OF STEREOCHEMISTRY

It has been observed that more than 100 years ago, the foundations of stereochemistry were laid down. Van't Hoff and Le Bel published their revolutionary papers about stereoisomerism and the tetrahedral orientation of the bonds of carbon in the year in 1874.

The focus on stereochemistry was not given until the early 1950's when the experimental verification of the theories laid for stereochemistry was fully supported by underlying theories of organic and basic chemistry. The field of stereochemistry flourished the attention that made it distinct or become a separate area of investigation. Thus, understanding stereochemistry requires adequate knowledge on the basics of organic, general, and biochemistry. Stereochemical thinking has become very important in both preparative organic chemistry as well as the mechanistic organic chemistry, in inorganic, in biochemistry, and polymer chemistry and other allied chemistry fields.

The extensive development of concepts and methodology took place in the period of 1960s and the early period of 1970s. The current past has been marked by a spectacular growth in the application of instrumental techniques—particularly of nuclear magnetic resonance (NMR) and chiroptical methods. These methods, moreover was achieved by the development of a consistent system of concepts as well as notations. These concepts and notions are based on symmetry principles and mathematical group theory.

On the other hand, it was marked by

It has been observed that in the last ten years, there were several Nobel prizes that have been awarded (to O. Hassel, D. H. R. Barton, V. Prelog, R. B. Woodward, and J. W. Cornforth) for accomplishments that are partly or completely contributed by scientists in the field of stereochemistry. These contributions were realized by the use of the modern system of concepts as well as notations that helped in honing the understanding and systematization of the phenomenon of stereochemistry.

In the present times, there are a wide range of topics that are covered by stereochemistry and also which was an expansion of the principles developed within the period 1950–1980. A big leap forward was introduced by Barton's discovery of the principles of conformational analysis (in the year 1950).

All these generalizations were warmly welcomed. This is because organic chemistry was just at a point in its growth where there was a huge requirement to properly understand the factors that have an impact on the conformational behavior of molecules.

Also, the availability of experimental and empirical information required general correlation and, after must have possible high rates of progress to be achieved in a number of directions. There was the formation of new areas of interest.

One of the most spectacular achievements in conformational analysis was the interpretations which it provided on the stereo-electronic aspects of chemical change. It becomes possible to propose reliable rationalizations of many features of chemical reactions which were shown by the alicyclic compounds.

The initial achievement by Barton in properly understanding the reactions of cyclohexane derivatives was extended rapidly to involve all the small, medium, and large membered carbocycles and heterocycles.

The interaction between conformational analysis and the studies that were done in the previous times by Hughes and Ingold (1935–1953) on the mechanism of nucleophilic aliphatic substitution reactions was responsible for most of the significant developments. These important developments have taken place later in the branch of physical organic chemistry.

Now, it has been observed that the analysis of the stereochemistry of organic reactions is considered as a recognized approach in the examination of the mechanism of organic reactions, in general. The principles of conformational analysis, in connection with an appreciation of the stereo-electronic aspects of reaction mechanisms, have had a great impact on the design of new synthetic methods. These methods are either stereoselective or stereospecific.

In the present times, effective control of the substance stereochemistry total synthesis of compounds that are having complex acyclic molecules or cyclic chiral molecules, is a recognized aspect of modern organic chemistry. In addition, there have been great advancements made in developing a well-based understanding of the phenomenon of asymmetric induction.

Woodward and Hoffmann around 1970 were able to correlate a huge array of chemical information on the basis of the simple concept. According to this simple concept, the concerted pericyclic reactions can proceed under favorable conditions only if the orbital symmetry is conserved without the imposition of unacceptable stereochemical constraints.

The structure of molecules can be considered in three levels. These three levels are known as the constitution, configuration, and conformation of a molecule. The initial direct determination of an absolute configuration was made by Bijvoet in the year of 1951. Following this, great advancements has been made in the determination of the absolute configurations of most of the chiral organic molecules either by direct methods or by correlative methods.

There has been a great interest in the stereochemistry of biosynthetic processes. This is because of the study of the absolute configuration of most of the natural products. In addition to it, there have also been some exciting as well as essential improvements in the definition of the stereochemistry of enzyme action. It can be truly said that in the present times, stereochemistry provides a vital connection between bio-organic chemistry, chemistry, biochemistry, and molecular biology.

In the synthesis of drugs, the molecular designs of the drugs are based on the accurate stereochemical detail. It has been observed that in the recent years there have been some primary changes in stereochemical concepts and also, there were some changes in the nomenclature.

The necessity to identify molecular chirality produced the excellent proposals of Ingold, Cahn, and Prelog (1966). The new language was also invented. This has allowed stereo chemists to define molecules and intermolecular reactions in accurate as well as in unambiguous terms.

Later on, the idea of stereochemical nonequivalence (that is heterotopicity) of the ligands, that are identical, has attained a fundamental significance in most of the enzymatic reactions. And, also, there was the removal of much of the mystery that was related with the stereospecificity of biological processes. In addition to it, similar progress have taken place in organic chemistry since the explanation of stereochemical entities in topological terms was advocated by Prelog in the year 1969.

As a result of this approach, there has been a fascinating development of chemical topology. The availability of up-to-date or modern instrumentation currently has also had a significant impact on the development of stereochemistry.

In the present times, the information or knowledge about the geometry of molecules in the solid-state is available rapidly by the use of automated X-ray methods. In addition to it, quantitative information on the conformational behavior of molecules in solution is also provided by the study of the temperature dependence of NMR spectra.

1.3. ISOMERS AND STEREOCHEMISTRY

Isomers are defined as compounds with the same molecular formula but differ in structure. Isomers can be further classified as structural or constitutional and stereoisomers. Structural isomers are substances (with the same molecular formula) but differ in the arrangement or placement of atoms in the compound, which will make their condensed structure different from each other. The changes connectivity of the carbon atoms is also one of the distinct features of constitutional isomers.

In stereoisomers, the atoms are connected sequentially in the same way, such that condensed formulas for the two molecules are identical. The isomers differ, however, in the way the atoms are arranged in space. There are two major sub-classes of stereoisomer; conformational isomers, which interconvert through rotations around single bonds, and configurational isomers, which are not readily interconvertible.

Chiral compounds, on the other hand, are defined as the compounds which are having the same molecular and condensed structural formula, but vary in the orientation of four distinct groups that are around the atom of carbon.

These are chiral compounds and geometric isomers or isomeric compounds that are present in the orientation of atoms on a double bond or ring system. Often, they are referred as stereoisomers since they belong to both categories (Figure 1.3).

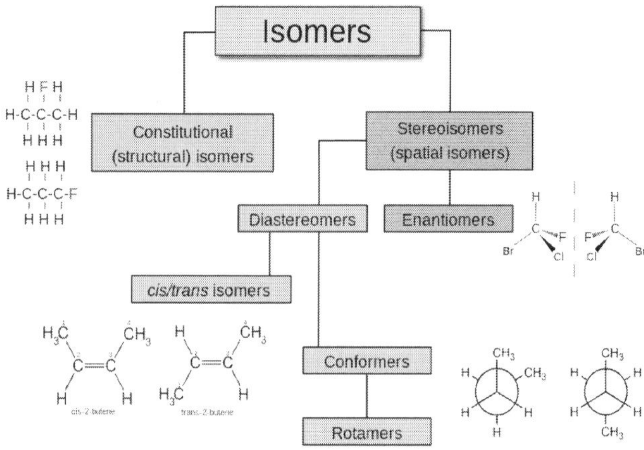

Figure 1.3. Classification of isomer in stereochemistry.

Source: Image by Wikimedia commons.

Conformational analysis studies about the different orientations that can be achieved by the molecules and the way their energy changes as their atoms are rotated in space, therefore, becomes important since these will draw the distinction between one molecule over the other. It has to be noted that conformations do not fall under the classification of isomers. This is because different conformations do not mean that they are different structures. Basically, they are the same molecule, only the bonds have rotated in the space.

1.3.1. Isomers

In isomers, it is important to remember that the only guaranteed similarity among them is their molecular formula. Different chemical structures can be separated from each other as they have different characteristics and properties such as their boiling and melting point are different from each other. Isomeric substances automatically differ in this parameter due to the difference in the spatial arrangements.

It is essential to consider that isomerism is a property between a couple of particles or more than that, for example, an atom is an isomer of another particle and one particle can have distinctive isomeric connections to two unique atoms, for example, A will be an established isomer of X and stereoisomer of Y.

Comparable connections happen inside families, for instance, you must be a sibling to another person while an individual can be both a sibling and a child to various individuals. Depending upon the idea of the contrast between the structures, it is highly likely to characterize the isomeric relationship into different sub-types.

Isomers are various mix of compounds with the equivalent sub-atomic equation yet, in some way, that are fundamentally different. It is essential to be able to recognize isomers since they can have distinctive compounds, biological properties, and physical properties.

1.3.1.1. Constitutional Isomers

As discussed earlier, constitutional isomers differ by spatial arrangements like in the connectivity of the carbon atoms or the hydrocarbon backbone. Figure 1.4 illustrates the isomeric compounds ethanol and dimethyl ether. Looking into the connectivity of carbon atoms, ethanol has the -C-C-O connectivity while dimethyl ether has the -C-O-C-hydrocarbon backbone. Table 1.1 further expounds the set of examples by adding another set of constitutional isomers in the form of butane and 2-methylpropane. Table 1.1, moreover presents the difference in the properties of each isomer.

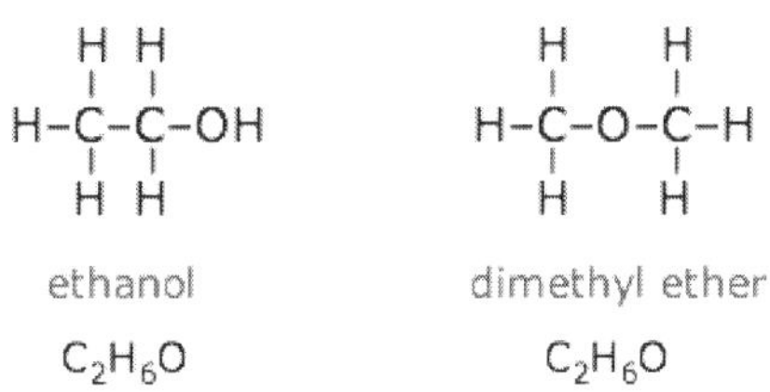

Figure 1.4. Ethanol and dimethyl ether are constitutional isomers.

Table 1.1. Illustrative Examples of Constitutional Isomers and Their Varying Properties Based on Their Structure

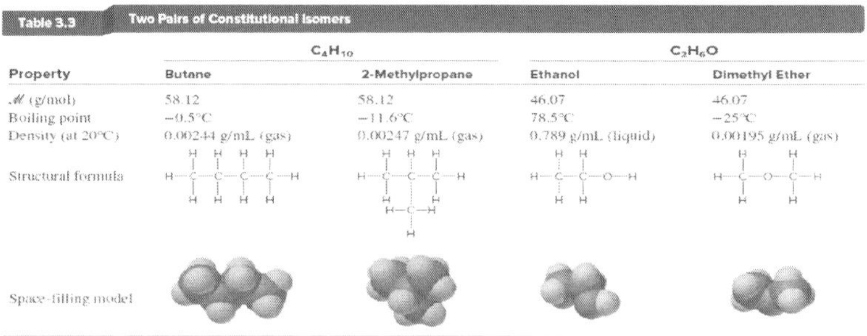

Table 3.3 Two Pairs of Constitutional Isomers

	C_4H_{10}		C_2H_6O	
Property	Butane	2-Methylpropane	Ethanol	Dimethyl Ether
$\mathcal{M}$ (g/mol)	58.12	58.12	46.07	46.07
Boiling point	−0.5°C	−11.6°C	78.5°C	−25°C
Density (at 20°C)	0.00244 g/mL (gas)	0.00247 g/mL (gas)	0.789 g/mL (liquid)	0.00195 g/mL (gas)
Structural formula				
Space-filling model				

In the case of butane and 2-methylpropane, notice that butane assumes the straight-chain configuration. This is because butane as illustrated is an aliphatic hydrocarbon while the methyl group attached to the 2nd carbon atom (whether you start the count form left or right) in 2-methylpropane classifies it now as a branched hydrocarbon.

1.3.1.2. Conformational Isomers

The Conformational isomers, also known as conformers or rotational isomers or rotamers. These are stereoisomers produced by rotation about s bonds and are often rapidly interconverting at room temperature [Figure 1.5 (i), (ii)].

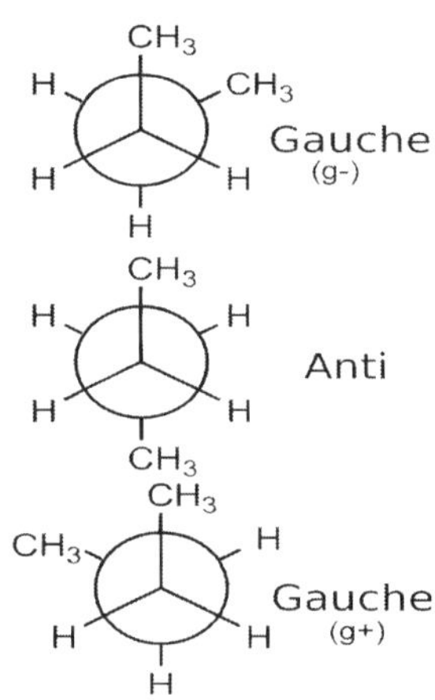

Figure 1.5. (i) Several examples of conformational isomers.

Source: Image by Wikimedia commons.

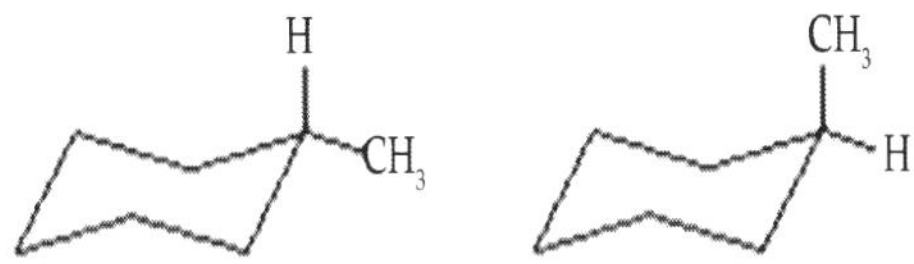

Figure 1.5. (ii) Another example of conformational isomers utilizing the chair conformation of methylcyclohexane.

1.3.1.3. Configurational Isomers

These are the stereoisomers that have some issues in interconverting and separation at the room temperature due to the fact that bond making and bond breaking are the processes required for their interconversion is not possible at room temperature. Configurational isomers, on the other hand, can be enantiomers or diastereomers (Figure 1.6).

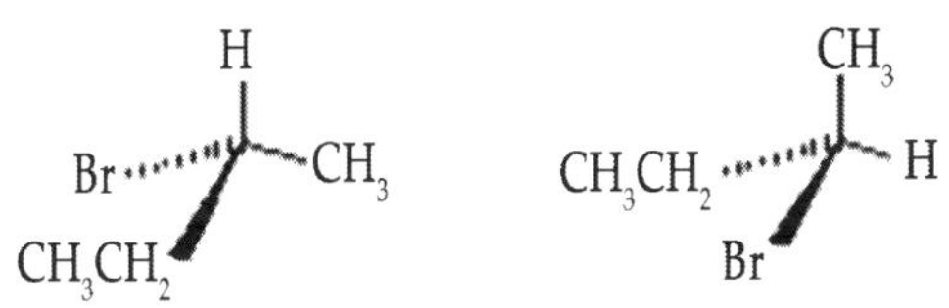

Figure 1.6. Structure of the configurational isomers of bromobutane.

1.3.1.4. Enantiomers

Enantiomers are stereoisomers which are related to each other as mirror images. Enantiomers can contain any number of stereogenic centers, as long as each center is the exact mirror image of the corresponding center in the other molecule.

With the help of this illustration, enantiomers can be easily understood through Figure 1.7. (R)-2-butanol is the mirror image of (S)-2-butanol. Thus, the compounds that are enantiomeric and are an identical representation of each other. However, they cannot superimpose each other and, hence, are non-superimposable.

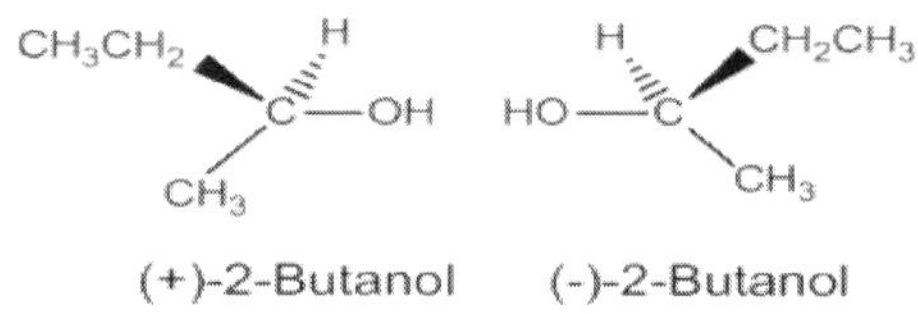

(S)-2-butanol (R)-2-butanol

Figure 1.7. *An illustration of enantiomers: Enantiomers of 2-butanol by S and R labels*

Source: Image by Flickr

It is also worth remembering that the stereogenic carbon's still retained its identity as the stereogenic centers of two enantiomers. The substituents attached with all the stereogenic carbon does not change in terms of its position and arrangements-what is at the back remains at the back and what is in front is still in front as well as those at the sides of the plane be still there. This is because they should be mirror images of each other.

1.3.1.5. Diastereomers

Diastereomers, on the other hand, display changes in the positioning and arrangement of the substituent of the stereogenic center. If one or more of these centers differs in configuration, the two molecules are no longer mirror images, but are totally different chemical compounds with differing physical and biological properties and are now considered to be diastereomers. Figure 1.8(i) represents diastereomerism through the changes in the locations of the substituents of the stereogenic carbons:

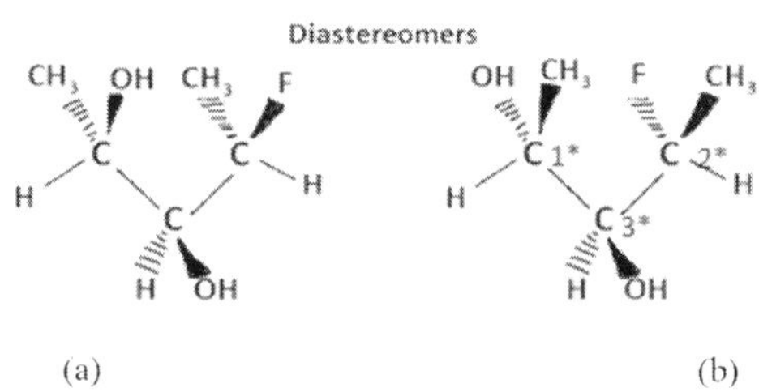

Figure 1.8. (i) *Diasteriomerism as illustrated in C1*, C2*, and C3*.*

Notice the changes in the C1* substituents where OH was initially in front and CH_3 is at the back of stereogenic carbon in molecule (a), looking into (b), the OH and CH_3 switched places putting CH_3 in front and OH at the back. The same trend can be observed in C2* and C3* locations from molecule (a) to (b). Since then, molecule (a) and (b) are not mirror images of each other because they are actually diastereomers and not enantiomers.

The number of diastereomers of a given molecule can be determined based on their stereogenic centers using the formula: $x = 2^n$; where x is the number of possible isomers and n is the number of stereogenic carbon or stereo centers. The molecule in Figure 1.8.(i), which obviously has three stereogenic centers (C1*, C2* and C3*), will have 8 isomers. Two of them is already presented, so you can practice and figure out the other 6.

Meso compounds are also classified as diastereomers. Though these compounds have two stereogenic centers, they have an internal plane of symmetry, which makes it an achiral molecule. The molecule can be split by an imaginary mirror so that all atoms on one side of the mirror are the exact reflection of the atoms on the other side. This can be seen below for the structures of *cis-(x, x)*-dimethyl cyclopentane; there are two chiral centers in the molecule since the two carbons are labeled with the red asterisk are each bonded to four different groups.

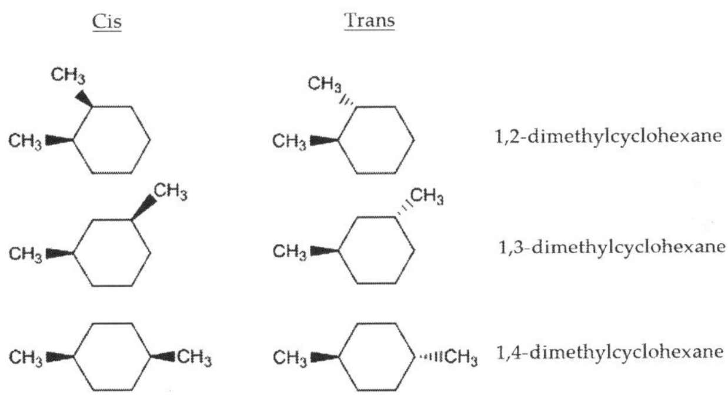

Figure 1.8. (ii) Examples of diastereomers: Dimethyl cyclohexane isomers.

Source: Image by Flickr.

Cis-and trans isomers are also known as geometric isomers. Geometric isomerism is visible among molecules where carbon is in single bonds and can have an unrestricted rotation in the stereogenic carbon. It is noteworthy to remember that multiple bonds (especially doubles bonds) cannot rotate.

1.4. PHENOMENON OF STEREOISOMERISM

Stereoisomers are non-indistinguishable atomic species having a similar constitution, for example, similar interconnections between their atoms. Contrasts between some stereoisomers might be communicated as far as interatomic separations, while others must be defined, for example, clockwise, and anticlockwise, left, and right, and so on.

Stereoisomerism is not restricted to three measurements since planar atoms, for example, 1,2-dichloroethene, may shape stereoisomers similarly. There are basically two distinct ways to deal with the conversion of stereoisomers.

The first considers only the symmetry of models under discussion and for this reason, it is the simpler of the two approaches. Right now, the quick geometry of a molecule, as represented by a stereo drawing or sub-atomic model, is significant, without considering the plausible lifetime of the specific state.

The subsequent methodology considers the vital boundaries which separate stereoisomers, and consequently makes presumptions with respect to the conceivable presence of these isomers.

At the high temperatures utilized by catalytic reactors considered for the isomerization of alkanes in treatment facilities, it is beyond imagination to expect to recognize even between the two protected isomers of butane, for example, n-butane and isobutane, in light of their quick interconversion.

The conversion of n-butane into isomeric configuration required 13 kJ/mol of energy to take place, near absolute zero temperature and must be maintained. A further representation of the vagueness of characterizing stereoisomerism based on the distinguishableness of isomers is given by some biphenyls which display impeded axes.

Since the pair of chiral biphenyls is converted by supplying of 125 kJ/ mol of energy, they can be isolated at room temperature, analyzed independently, and their stereoisomer nature, in this way, is preserved.

At 200°C on account of their quick interconversion, it is not, at this point conceivable to isolate the stereoisomers. With the biphenyls rotation is reduced (75 kJ/mol), as a result, interconversion is quick to the point that preparation is not possible even at room temperature.

In this way, if distinguishableness is taken as an essential aspect for stereoisomerism, the biphenyls stops isomerism at higher temperatures while the biphenyls are not stereo-isomeric even at room temperature.

It ought to be noticed that stereoisomers which might be indistinguishable from the preparative perspective, can be recognized through several spectroscopy techniques (where NMR spectroscopy is most common)

Clearly, if the practical separation of stereoisomers is being considered, then the record is ought to be taken off of the barriers which separate them. However, before this, it is essential to inspect them from the perspective of balance alone.

This allows the proper inspection of the stereoisomerism among the molecules considering proper separation aspect.

1.4.1. Chirality and Optical Activity

Chirality is primarily associated to the presence of chiral centers in the molecule. Chiral centers tend to be the point of rotation of molecules and therefore, chirality is also associated with the property of a molecule to rotate the plane of polarized light. An atom that is attached to four distinct substituents in a nonplanar arrangement is known as a chiral center or it can also be called as stereogenic center or stereocenter asymmetric center.

Optical activity is the potential of a chiral molecule to rotate the plane of the plane-polarized light. t can be measured or gauged with the help of a polarimeter. A simple polarimeter is consisting of a light source, polarizing lens, and sample tube along with the analyzing lens.

When the light goes through a sample that can rotate the plane-polarized light, the light appears to dim because it no longer passes straight across the polarizing filters. The amount of the rotation is quantified as the number of degrees that the analyzing lens must be rotated in such a way that it appears as, there is no dimming of light has taken place.

An is *optically active* substance is one which can *rotate the plane of polarized light*. In a pair of optically active enantiomers, each enantiomer will rotate the plane of polarized light in equal and opposite directions; these enantiomers can therefore be referred to as (+) and (–) enantiomers, according to the direction of the observed rotation. In some cases, there will be equal concentrations of both enantiomers is produced and will display no net optical rotation. These are therefore called "racemic mixtures" (Figure 1.9).

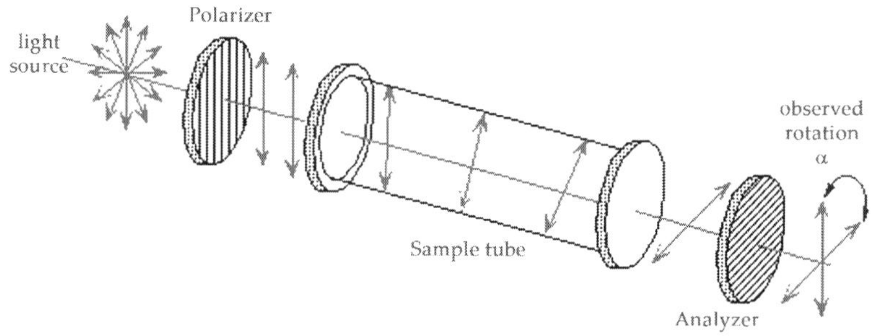

Figure 1.9. Schematic diagram of a polarimeter.

The angle of rotation is measured in a device known as a *polarimeter*. The substance in question is typically dissolved in a solvent at a known concentration (***c*** in grams per mL) and placed in an analysis tube of a known length (***l*** given in decimeters; one dm = 10 cm). The rotation observed in the polarimeter is denoted with the Greek letter α, and the *specific rotation* for a molecule (denoted as $[\alpha]_D^{25}$) is given by the equation shown below:

$$[\alpha]_D^{25} = \frac{\alpha_{obs}}{c * l}$$

The specific rotation for a molecule is also dependent on the wavelength of the plane-polarized light. A common light source for simple polarimeters is a lamp with enhanced output at the "sodium D-line"; in this instance.

Problem 1:

A pure enantiomer has an observed optical rotation of –0.82° when measured in a 1 dm tube at a concentration of 0.025 g/mL. Calculate the specific rotation for this molecule.

Solution:

***c* = 0.025 g/mL; *l* = 1.0 dm**

The specific rotation is, therefore:

$[\alpha]_D^{25} = \frac{-0.82}{(0.025 * 1)} = -32.8^{\circ}$ g-ml-dm or -32.8° g^{-1} ml^{-1} dm^{-1}

If the specific rotation of a pure enantiomer is known, the observed rotation can also be used to calculate *optical purity*, or the level of contamination of one compound with its enantiomer, using the simple convention:

$$\text{Percent optical purity} = \frac{(\text{moles of one enantiomer-moles of the other enantiomer})}{(\text{moles of one enantiomer+moles of the other enantiomer})} * 100\%$$

$$= \frac{\text{observed specific optical rotation}}{\text{specific optical rotation of the pure enantiomer}} * 100\%$$

Problem 2:

The observed specific optical rotation of a compound is [α]= +5.00°. The specific optical rotation for the pure enantiomer is:

$$\text{Percent purity} = \frac{+5}{+20} X100 = 25\%$$

The sample consists of 75% of the racemic form (=equimolar mixture of both enantiomers, α=0°) and an excess of 25% of the enantiomer in question (62.5% and 37.5%).

1. **Optical Isomerism:** Optical isomers are named so because of their effect on plane polarized light. (Two) Enantiomers are simple substances which show optical isomerism.

- The enantiomer that carries the (+) is the solution of one enantiomer that rotates the plane of polarization in a clockwise direction.
- Thus, for example, in the enantiomers (+)-2-butanol and (–)-2-butanol, the first rotates the plane of polarization in the clockwise direction while the latter rotates it in the counterclockwise direction.
- If the solutions are equally concentrated the amount of rotation caused by the two isomers is exactly the same but in opposite directions.
- When optically active substances are synthesized in the lab, they often occur as a 50/50 mixture of the two enantiomers. This is known as a *racemic mixture* or *racemate* and has no effect on plane-polarized light.
- An entire molecule is divided into two by an imaginary plane of symmetry. The resulting molecules are exact mirror images. When the resulting molecules displayed the same mirror image of the other one, the molecule "has a plane of symmetry" and that molecule is "achiral." Conversely, molecules that lack plain of symmetry are non-superimposable with each other and are classified to be chiral molecules.

Therefore, enantiomers are chiral molecules. A molecule, like lactic acid, that is having a tetrahedral carbon with four distinct attachments may also exist as a pair of enantiomers. Also, between the two enantiomers, there exists no degree of rotation that will allow the molecule to interconvert.

It has been observed that tetrahedral carbons are the mostly encountered centers of chirality. Chirality is not just restricted to the atoms of carbon alone. Atoms of phosphorus, nitrogen, and silicon can also form tetrahedral structures and thus, are potentially chiral in nature.

It has to be noted that the non-bonded lone pair of electrons acts as a substituent in the trigonal pyramidal structure of the compounds of phosphorus and nitrogen. Therefore, unsymmetrical phosphines have a possibility of being chiral at room temperature. On the other hand, no amines can show chirality.

If it is assumed that the lone pair, possessed by the trivalent nitrogen, is considered as a fourth substituent then the trivalent nitrogen can be considered as tetrahedral. At nitrogen, the pyramidal geometry does not make a chiral center since it flips rapidly, and the enantiomers equilibrate too quickly to be isolated. Therefore, the unsymmetrical tetrahedral nitrogen molecules are achiral in nature.

On the other hand, trivalent phosphorus flips more slowly as compared to that of their nitrogen counterpart. Therefore, the isolation of individual enantiomers of phosphines is practically possible. Therefore, the unsymmetrically substituted trivalent phosphorous compounds are chiral in nature (Figure 1.10).

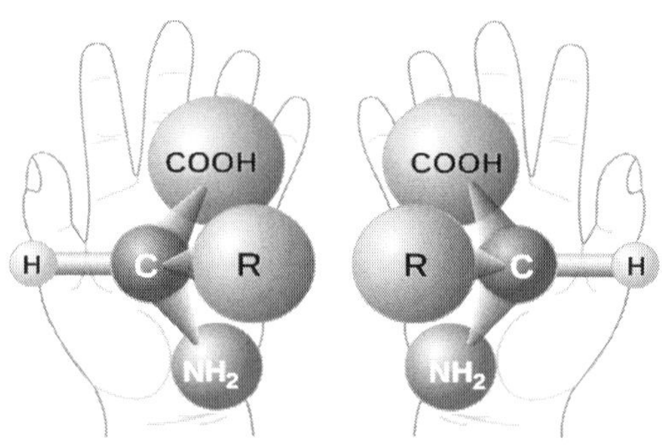

Figure 1.10. An illustration of chirality using hands.

Source: Image by Wikimedia commons.

Silicon also makes tetravalent compounds apart from carbon. Therefore, the unsymmetrically substituted silicon compounds are chiral in nature. Silicon compounds that are chiral in nature, find various applications in synthesis as well as in chiral chromatography.

All those metal complexes that have tetrahedral geometries or octahedral geometries might also be chiral in nature if they are unsymmetrical. Tetrahedral complexes, which are similar to the structure of carbon bonded to four distinct groups, are chiral in nature. Therefore, these exist as enantiomers (optical isomers). The tetrahedral $[Co(NH_3)ClBrI]^-$ ion, that is having four different groups complexed to cobalt has two non-superimposable mirror images. It has to be noted that molecules which are chiral in nature may have more than one chiral center.

1.4.2. Chirality Center

A molecule needs the presence of a chirality center or a stereogenic center to exist as a pair of enantiomers. Several terms like an asymmetric, stereogenic or chiral center have been used in the previous times but in the present times, these terms have been replaced by the term chirality center. This is done in order to eradicate the ambiguity (Figure 1.1).

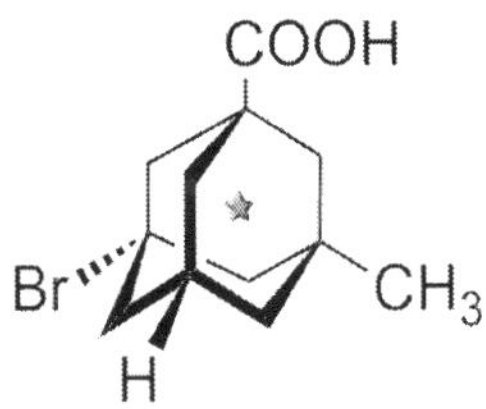

Figure 1.11. An example of chirality center: Adamantane derivative chirality.

Source: Image by Wikimedia Commons.

It has to be noted that the term stereocenter is also used to explain a point in the spatial arrangement of the molecule that leads to stereoisomerism. A molecule can have one or more stereocenter and a molecule with more than one stereocenter is a chiral molecule. A chirality center is characterized by an atom with four different groups that are bonded to it in such a way that it has a non-superimposable mirror image.

If a single chirality center is present in a molecule, it results in a chiral molecule. This is because there is a different bottom, top, right, left, front, and back faces. Of course, a molecule can possess more than just a single chirality center. It has to be noted that such kinds of molecules may or may not be chiral in nature.

1.5. STEREOISOMERISM IN PENTA-AND HEXA-CO-ORDINATE COMPOUNDS

There are various reactions that require pentacoordinate transition states of phosphorus and arsenic atoms. Phosphorus and arsenic atoms attain 5-coordinate transitions states in some reactions while in other reactions, their compounds and several reactions require 4-coordinate transition states,

Some of the reactions that require 5-coordinate transition states are the hydrolysis of phosphates. On the other hand, the hexa-coordinate transition state is rather common in the transition-metal complexes.

In all the complexes where the extent of coordination exceeds four, such as the penta-coordinate and the hexa-coordinate transition states, stereoisomerism becomes quite complex.

This is not at all surprising when one considers that even in the most symmetrical arrangement of the ligands in 5-coordinate compounds, i.e., in the trigonal-bipyramid of D3h symmetry, it is possible to distinguish two geometrically different situations (Figure 1.2).

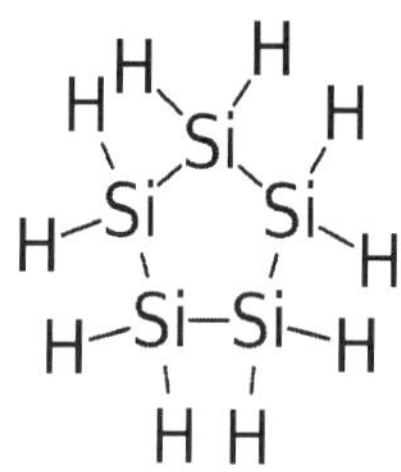

Figure 1.12. An example of penta-compound: Pentalsilolane.

Source: Image by Wikipedia.

Therefore, two of the ligands are situated at the ends of the C3 axis (that is in apical positions). On the other hand, the three ligands that are remaining may lie along the C2 axes. These ligands lie in equatorial positions.

Thus, it follows that with Xaaaab-type substitution, there exists two stereoisomers. One stereoisomer in which the ligand V is in an apical position and the other ligand 'b' being equatorial. It has to be noted that if the 5-coordinate compound is acyclic in nature, chirality is acquired when there are four different ligands (Xaabcd) present. In this, one 'a' ligand is in an equatorial position. And the other 'a' ligand is in another position that is apical position (Figure 1.13).

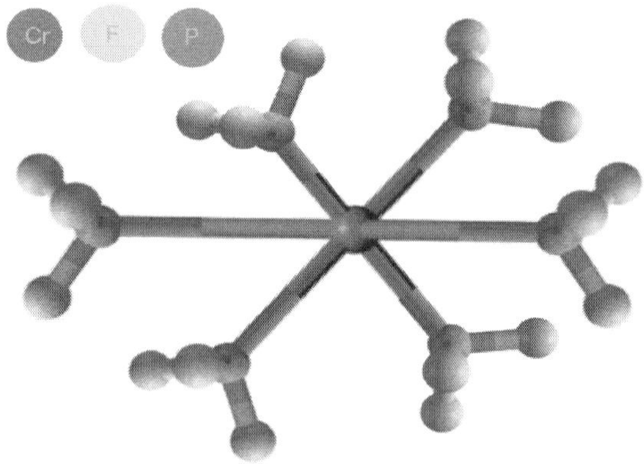

Figure 1.13. An example of hexa-compound: Chromium hexa-(trifluoro phosphine).

Source: Image by Wikimedia commons.

Although, when all the ligands are different in nature than every one of the 20 stereoisomers, the molecule is chiral in nature. One should keep in mind that with 4-coordinate compounds, the process of diastereoisomerism is not possible when there exists just a single chiral center.

On the other hand, in 6-coordinate compounds, the ligands place themselves on the apices of an octahedron. It must be noticed that all of the apices of the octahedron are geometrically equivalent.

The only important information that needs to be mentioned here is that the position of two ligands. If the ligands are equivalent, they may occupy adjacent or opposite apices of the octahedron to form the cis-and trans-diastereomers, respectively.

1.6. ASYMMETRIC SYNTHESES

The process of synthesis of enantiomerically pure organic compounds poses a challenge to both the academic chemist as well as the industrial chemist in the same way. It has been observed that the art of organic synthesis has made it to a certain point where it is completely conceivable to make by total synthesis many sophisticated compounds, which are especially useful for their biological properties.

There are a variety of other agents such as L-dopa. It is an active agent against Parkinson's disease, while amino acid D-penicillamine has indicated some favorable effects on the clinical features of rheumatoid arthritis (Figure 1.14).

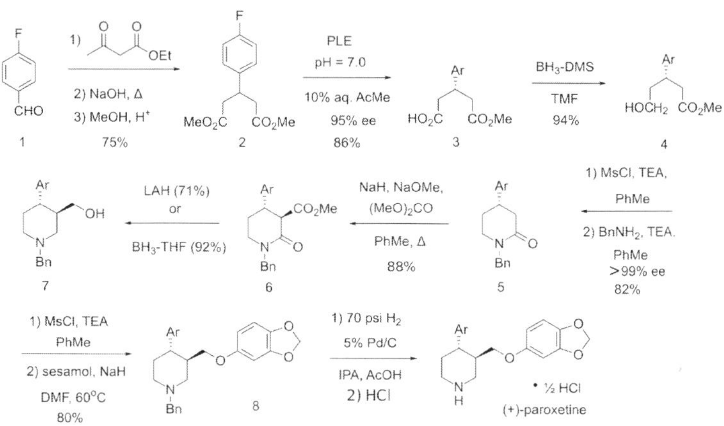

Figure 1.14. An example of asymmetric synthesis: Paroxetine asymmetric synthesis.

Source: Image by Wikimedia commons.

It has been observed that there is one or more than one asymmetric center in most of the biologically active compounds and the particular activity is being related just to a single stereoisomer. For example, the drug that is having (S, S)-configuration is nearly 100 times more active as compared to the corresponding diastereomer that is having (R)-configuration in the side chain.

Some other examples for molecules that exhibit different biological activities of the stereoisomers consist of pheromones, herbicides, and many more. There are several numbers of drugs that are made by the process of total synthesis. These drugs contain at least one asymmetric center. But it has been observed that just 20% of these drugs have been used in sterically pure form.

In this way, the primary necessity is to be able to regulate the formation of the asymmetric centers through the course of the process of synthesis.

More particularly, delicate is the formation of the first asymmetric center in an achiral molecule. The tactics or approaches of asymmetric synthesis is to find out the solution of this issue.

The term "asymmetric synthesis" is first used in the year of 1984, by the well-known researcher named E. Fischer. In addition to this, in the year of 1904 is defined by the well-known researcher named as W. Markwald.

W. Markwald gives the definition of asymmetric synthesis as, "it is a reaction that develops optically active substance from symmetrically constituted compounds with the intermediate application of the optically active materials but with the exclusion of all analytical processes."

After that, a modern definition was proposed by Morrison and Mosher stating that, "an asymmetric synthesis is a reaction in which an achiral unit is an ensemble of substrate molecules is transformed with the help of reactant into a chiral unit, in such a manner that, the stereoisomeric products (such as enantiomeric or diastereomeric) are formed in unequal amounts.

In this way, an asymmetric synthesis is a process, that transforms a prochiral units into a chiral unit, in such a way that, as an outcome, formation of unequal amounts of stereo isomeric products takes place. Never the less, a significant effort has been put forward with the help of chemists in order to accomplish comparable results. It is quite a difficult task in order to form chemical system as effective as enzymic ones.

For several numbers of years, it has been argued that, whether optical yields could be efficiently achieved or accomplished with the help of chemists without the help of any sort of enzymes. Nevertheless, an increment in the amounts of latest outcomes shows that the versatile as well as effective non-enzymatic asymmetric synthesis are certainly achievable.

The ever-increasing knowledge, with respect to the approaches or methodologies of asymmetric synthesis is demonstrated with the help of several numbers of reviews that have appeared in the literature. More particularly, the latest ones by Morrison and Mosher, Scott, and Valentine, Meyer, Kagan, and Fiaud, Otsuka, and Tani, Apsimon, and Sequin, Fischli, Wynberg, Masamune, and Choy and Heathcock.

Nevertheless, these several numbers of reviews did not cover in depth the latest development or improvement that took place in the asymmetric synthesis with the help of chiral amino acid reagents. The enantiomeric products of the asymmetric synthesis-the chiral units can either be prepared in the laboratories or be isolated from natural products.

In this way, several chemists have at his disposal a huge number of chiral units among the natural products. More especially, carbohydrates, terpenes, t-hydroxycarboxylic acids, alkaloids, biogenic amines, and amino acids have found application as building blocks for asymmetric syntheses.

On the other hand, there is also a disadvantage in the application of the carbohydrates, alkaloids, and terpenes, which can hardly ever be attained in both of the enantiomeric forms, does not exist in the case of hydroxycarboxylic acids such as lactic, mandelic, malic, and tartaric acid.

1.6.1. Heterogeneous Asymmetric Hydrogenation

When nickel, along with the other transition metal catalysts are "modified" with a chiral compound such as (R, R)-tartaric acid, they become enantioselective. All efforts, in order to modify the solid surfaces with optically active substances, have so far brought about catalysts of only low stereoselectivity.

And the reason behind this is that, there are several numbers of centers that exist on the surface of the catalyst. In this way, with respect to asymmetric hydrogenations, the methodologies of homogeneous catalysis are superior with respect to the heterogeneous catalysis.

Nevertheless, several numbers of carbonyl compounds have been hydrogenated in the presence of the tartaric-acid-supported nickel catalysts in up to 92% optical purity.

1.7. STATIC STEREOCHEMISTRY

The several numbers of factors of the stereochemistry which are the properties of the ground state of molecules will be discussed in this chapter. The methodologies or tactics, with the help of several molecular species that interconvert will be addressed further in this section.

In order to differentiate among the several numbers of various molecular species, chemists, and more particularly organic chemists, use terminologies which are having different meanings, different significance and emphasis.

The compounds that are having similar molecular formulae can differ with respect to their chemical properties, that has been well acknowledged since the early 1800s. One of the fundamental accomplishments is the early attention and focus on constitutional (or structural) isomerism.

Variances or fluctuations in the composition or the bonding connectivity can also be demonstrated with the help of two-dimensional formulae without recourse being necessary with respect to the expressions that are navigating with the direction such as left, right, below, above, etc.

For example, the two alcohols with the same molecular formula of C_8H_8O, can be differentiated since in 1-propanol, the oxygen atom is linked to a primary carbon, usually the single atom of carbon on either side of the structure. On the other hand, in the context of isopropyl alcohol or 2-propanol, the oxygen atom is linked to a secondary carbon which is the second carbon atom, in the life-wise or right-wise direction.

The concern of stereochemistry is the concern regarding isomers beyond the position or location issues of the substituent's, and the variations that cannot be expressed with respect to the atomic interconnection. In this way, two different concepts are very significant in the context of stereoisomerism- the molecular configuration and conformation.

Talking about molecular configuration, it is essentially a qualitative term which explains the relative arrangement of the atoms of a particular given molecule in space. A single molecule can assume several configurations which highlight the different arrangement of atoms the comprises such molecule.

Various conformation of the molecules which are having the same composition as well as same configuration fluctuate with the help of their angle of torsion around single bonds. Conformational changes may, or may not, impact the configuration.

1.8. PLANAR CHIRALITY

Chirality, is a topological property of the handedness of a given structure of a particular molecule. It is known to be one of the most interesting aspect or characteristics of a molecule that also serves as the backbone of stereochemistry. The term, chirality, was first coined by a Lord Kelvin, in the year of 1886 and after that, was made popular onto chemistry by the year of 1964 onwards. It was soon accepted all across the globe (e.g., in the year of 1965, in the field of metallocene chemistry), to replace the old and mostly inappropriate used terms asymmetry and dissymmetry and has been then the subject of several numbers of various discussions, interpretations, considerations, and factorizations.

Moreover, to the wide-ranging and long known field of compounds that are mainly made up with the help of chiral substances since the studies on the axis and the plane of chirality were introduced in the year 1956 to distinguish chiral substances and their difference in the number of l compounds that e lacks a chiral center.

This further sets the three stereogenic units as the basis of the generally accepted CIP (which is the short form of Cahn-Ingold-Prelog). On the other hand, a center or an axis of the chirality can be clearly explained, but there is still some uncertainty or doubt with respect to the specification or the characteristics of a plane of the chirality.

In this way, it seems quite a challenging task, to define the scope of the planar chiral compounds which at a first glance were rigid aromatic compounds of the special interest from the synthetic, structural, spectroscopic, and more particularly chiroptical points of view.

The first definition of a chiral plane (plane of chirality) was proposed in the year of 1956, as " the result when a plane of symmetry is destroyed in such a way that chirality arises only with the help of the variations of both sides of the plane" (that is, a plane of chirality distinguishes two dimensions from the third one) and is appropriate for the chiral (and optically active) compounds known at this time which are not covered by the other two stereogenic units (which are center as well as axis of chirality). These stereogenic units are the cyclophanes–or the so-called Ansa and/or the Para compounds, which are also known as cyclophane derivatives.

Nevertheless, it does not explain enough for the categorization of other organic structures, for example, the large group of chiral metallocene (more particularly, ferrocenes) and maybe, also for bridged and [14]-annulenes and (E)-cyclooctene, which are not planar with respect to the above definition.

In this way, (E)-cyclooctene and alkenes that are related to that compound, could, and were also regarded as axial chiral, an elated [2.2] meta cyclophanes also helical.

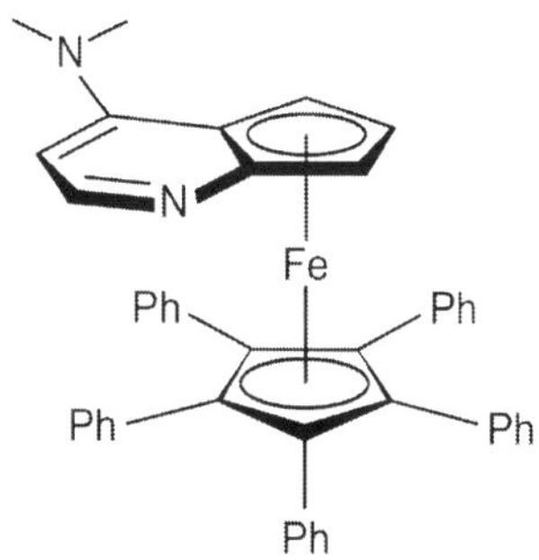

Figure 1.15. An example of planar chirality: Planar chiral ferrocene derivative.

Source: Image by Wikimedia Commons.

Unsymmetrically disubstituted (which is a chiral structure) metallocene (with point group C_1), for example, the ferrocene I 2z, provide a rather special issue. These sorts of compounds were first defined as planar chiral and the characteristics or the properties of the molecular chirality (descriptors Rp and Sp) applied accordingly. Several scholars up to now, still categorize these structures as planar chiral.

Nevertheless, Cahn, Ingold, and Prelog (CIP) proposed a system or organizing how chiral molecules and stereoisomers are named according to the chiral centers and the double bonds or multiple bonds present in a given organic molecule structure. This established the delineation between isomers and the formalism in identifying and distinguishing one isomer from the other.

This formalism has taken the direction towards a new definition for chiral planes which is also consisting of metallocene. In order to evade these kinds of issues, the term "metallocene chirality" is referred to with the help of, nevertheless, the descriptors (R) and (S) as specified for centro-chirality.

Metallocenes are compounds typically having two cyclopentadienyl anions (C_5H_5 , abbreviated as Cp) attached to a metal center (M) in the oxidation state II, and follows a general formula $(C_5H_5)_2M$. On the other hand, metallocene derivatives are substances known to be closely related to a metallocene. Examples are; titanocene dichloride, $(\eta^5\text{-}C_5H_5)_2TiCl_2$, commonly abbreviated as Cp_2TiC, which is a common reagent in organometallic and organic synthesis and vanadocene dichloride, an organometallic complex with formula $(\eta^5\text{-}C_5H_5)_2VCl$ (abbreviated as Cp_2VCl_2)-is a structural analogue of titanocene dichloride but with vanadium (IV) instead of titanium

(IV) as the metal center. Majority of the metallocene and their derivatives have catalytic properties, but are rarely used industrially.

Metallocenes are a subset of a broader class of compounds called sandwich compounds. In the structure shown in Figure 1.15, the two pentagons are the cyclopentadienyl anions with circles inside them indicating they are aromatically stabilized.

1.9. APPLIED STEREOCHEMISTRY

The potential which is required to allow the formation of stereoisomers is one of the most important in organic chemistry. The stereochemical reactions which progresses to give the results or the end product relies on the overall symmetry of the system, which is consisting of the substrate, the reagent, the medium, and not least of any catalyst used.

There are two possibilities that exist themselves with respect to this mechanism. The first possibility is the symmetry of the substrate and that of the reagent enables the formation of only one stereoisomer. The second possibility is one or more stereoisomer can be spontaneously formed.

The reverse of the second possibility would be that of stereo destructive reactions, such as transformations, which decreases the number of possible stereoisomers. In simple language, it transforms two or more stereoisomers to the same product.

Stereochemical changes that are linked with chemical reactions may also take place in either the transformation or replacement of ligands at centers, or alternatively a shift or modification in the coordination number of the atoms involved.

With respect to reason, it is, therefore, suitable to evaluate the formation of stereoisomers either with respect to the intramolecular symmetry of groups or with respect to the molecular faces.

Groups that are structurally similar, are those groups which have the same composition, components, structure, configuration as well as the equivalent positions. The intramolecular symmetry of these groups has already been discussed earlier in the chapter. The connection between their topicity and the stereochemical structure as well as the relative rates of their transformations is considered.

In the meantime, the homotopic groups behave in a strictly identical manner in any type of interaction. The homotopic groups react in the same manner with respect to any reagent (also with the biological ones) under all situations or cases. In simpler words, the homotopic groups develop or form the identical stereoisomer at the same reaction rate irrespective of which individual group in a homotopic set is involved.

The result of the reactions is consisting of enantiotropic groups, which rely on the symmetry of the system. Enantiotropic groups react under the achiral conditions, for example, with an achiral reagent, at the same rate, and in this way, the final product consists of a 1:1 (racemic) mixture of enantiomers.

Since the interaction of enantiotropic groups with an achiral environment is enantiomorphs in nature. In simpler words, it is similar in a scalar sense, and these kinds of groups cannot be differentiated under the achiral settings.

On the other hand, with respect to the chiral reagents, the fit with the substrate is different, and it relies on whether attack takes place at the pro-(R)-or 2 >pro-(£)-group, and in this way, the rate of the reaction with respect to the two groups will fluctuate since the activated complexes formed prior to the products will be diastereomeric and for this reason of different free energy.

If in the product, the reagent integrates with the substrate (e.g., with the help of substitution or addition), two diastereomers will be formed which are having different proportions (that is known as diastereoselectivity). In this way, the reaction of 3-phenyl glutaric acid with (iJ)–l-phenylethylamine takes to the formation of two diastereomeric monoamides, and they are having a ratio of 3: 2.

If only transient interaction takes place with a non-racemic chiral reagent (e.g., when the chiral reagent works as a reducing agent), the end product will be a non-racemic mixture of enantiomers (and this is known as enantioselectivity).

In addition to this, achiral reagents that are under chiral settings, more particularly in the presence of chiral catalysts, can also convert the substrates with enantiotropic groups into a nonracemic mixture of enantiomers.

Nevertheless, with a non-enzymatic process, enantioselectivity is generally low since the fit between substrate and reagent differs only slightly for the two groups. Taking the example of a process with relatively high enantioselectivity is provided with the help of the photocatalyzed isomerization of cis-l, 2-diphenylcyclopropane to its trans-isomer in the presence of a chiral sensitizer.

As the natural chiral catalysts, enzymes develop a very intimate as well as a highly orientated contact with the substrate and the reagent in the state of transition. As an outcome, the free energies of activation for reactions between enantiotropic groups fluctuates with a considerable amount and in this way, it is found almost invariably that only one of the two possible enantiomers are formed.

For instance, in the enzymatic phosphorylation of glycerol, the pro-(2)-hydroxy methyl group is completely esterified. The enzymatic oxidation of ethanol, which is the reverse process of the reduction in which one of the enantiotropic α-hydrogens is selectively detached, which is an example of an enantioselective stereo destructive process.

Diastereotopic groups are located in the chiral or achiral molecules that react at various speed of reactions with both chiral as well as achiral reagents and creates a diastereomer, which is in unequal proportions.

In this way, the diastereotopic α-methylene hydrogens in the sulfoxides exchange with deuterium at various rates in the presence of CD_3ONa in CD_3OD, with an exchange ratio for being 12:1 in favor of the eso-hydrogen.

The process of metalation of the dithiane is diastereo-, more particularly in the equatorial position, when the reaction is subsequently quenched with DC_1.

1.10. CONCLUSION

Stereochemistry is the branch of chemistry that focuses on the spatial arrangement of the atoms of the chemical elements that are involved in a chemical reaction.

Stereochemistry has some major aspects that involve the behavior of atoms that are placed at different angles. It is important to understand how these atoms behave differently when they are placed at different edges of various chemical compounds.

The most suitable examples of chemical compounds from which all major aspects of stereochemistry can be understood are the organic compounds. The organic compounds form different shapes under different conditions.

These different shapes are known as the stereoisomers. Major kind of stereoisomers are the enantiomers and the diastereomers. These are the types of isomers that are used to form the racemic mixtures. Racemic mixtures are achiral in nature and do not possess any kind of optical activity.

The organic molecules, when in contact with different chemical elements as their substrate, display different properties of chirality and their stereochemical activity changes. This makes the organic compounds with different chirality and different stereochemical activity useful for different reactions and are beneficial for the different industrial and medicinal purposes.

REFERENCES

1. (n.d.) Overview of Stereochemistry What is Stereochemistry? [ebook] uonbi.ac.ke. Available at: https://profiles.uonbi.ac.ke/andakala/files/sch_302_overview_of_stereochemistry.pdf (accessed on 18 August 2020).
2. Chem.ucalgary.ca. (n.d.). Ch 7: Stereochemistry. [online] Available at: http://www.chem.ucalgary.ca/courses/351/Carey5th/Ch07/ch7–0.html (accessed on 18 August 2020).
3. Cssac.unc.edu. (n.d.). Overview of Isomers and Stereochemistry — CSSAC. [online] Available at: http://cssac.unc.edu/programs/learning-center/Resources/Study/Guides/chemistry–261/overview-of-isomers-and-stereochemistry (accessed on 18 August 2020).

Chapter 2 Concepts and Terminology in Organic Chemistry

CONTENTS

This chapter discusses the various terms related to organic chemistry such as stereogenicity and stereoisomerism. The chapter also discusses the basic concepts and terminologies related to organic chemistry, stereochemistry, and organic reactions. The chapter also throws some light on topological and supramolecular stereochemistry in which structures such as molecular loops and knots, topological chirality, non-planar graphs, and achievements is discussing in relation to topological and supramolecular stereochemistry.

In addition, the chapter also explains topological junctive/disjunctive processes. Moreover, the ligogenic/lipolytic processes in organic chemistry is also tackled. The chapter also highlighted morpho selectivity and the relationship between morpholytic and morphogenic selectivity. The chapter also allots time in explaining chirality and stereoisomers together with the various related terms such as spatial arrangement, stereoisomers, enantiomers, chirality, compounds with multiple chiral centers being discussed. In the end, the chapter explains the stereo isograms of octahedral complexes.

2.1. INTRODUCTION

Elaborating further stereochemistry, it is also responsible for the explanation of the static aspects as well as the dynamic aspects of the three-dimensional shapes of molecules. This branch of chemistry has provided a foundation for understanding structure as well as the reactivity, particularly of organic molecules that are visually and spatially alike. At the same point of time, it has also been observed that stereochemistry also constitutes an intrinsically interesting research field in its own right.

Most of the chemists find this branch of study interesting simply due to the aesthetic beauty that is related with the chemical structures, and the intriguing capability of bringing together the areas of topology, geometry, and chemistry in the study and analysis of three-dimensional shapes.

In addition to it, there are some extremely significant practical ramifications of stereochemistry. It has to be noted that nature is inherently chiral. This is because the building blocks of life (that includes,-amino acids, sugars, and nucleotides) are chiral in nature and they appear in pure forms in nature.

Therefore, any of the substances that are made by humankind to interact with nature or modify nature by allowing it to interact with the environment that is chiral. This is considered as a vital concern for Bioorganic Chemists, while, it is considered as a practical issue for Pharmaceutical Chemists.

In the present times, Food and Drug Administration (FDA) needs that the drugs should be formed in enantiomerically pure forms. They also required that rigorous tests should be carried out in order to make sure that both enantiomers are secured.

In addition to it, it has been observed that stereochemistry is highly relevant to systems that are unnatural in nature. Also, all the properties of synthetic polymers are reliant on the stereochemistry of all repeating units. Ultimately, the study of this branch of chemistry, can be used in order to probe organic reaction mechanisms.

There are many research papers that provide the information about the fundamentals of stereoisomerism. In addition to it, there are many research papers or journals that have the information of the modern viewpoint, highlighting all the new terminologies, concepts, and techniques. This help students get a fundamental understanding of the basic principles of stereochemistry and all the terminology that is related to it.

2.1.1. Stereogenicity and Stereoisomerism

One of the reasons why stereochemistry is eternally challenging to scholars is because of the complex skill required in analyzing and visualizing three-dimensional objects, provided the two-dimensional representations on the paper. There are many models such as physical models and 3-D computer models that can be of great importance here, and these models must be used in order to have better understanding of the stereochemical reactions.

However, it has been observed that only simple wedges as well as dashes are given in almost all the drawings. These simple representations are the kind of representations that one must master. This is because usually, there is no availability of attractive, computer generated pictures at the work bench.

The familiar "wedge-and-dash" notation is the most common convention seen and used in understanding the stereochemistry of a molecule. It is important to note that there are some variations in the symbolism used in the literature. Usually, a dashed wedge grew in size gradually as transfers away from the point of attachment. This is applicable in case of receding group. Though, if one takes into consideration the ability of perspective drawing, there is no sense that the wedge gets bigger in size as it moves away.

Yet, this is the convention that is used most commonly. It has been observed that there are many workers who have turned to a simple dashed line instead, or a dash that decreases in size. In the similar way, both a bold wedge as well as a bold line are used in order to represent forward-projecting substituents. The other common convention is the bold "dot" on a carbon at a ring junction.

This represents a hydrogen that is projected toward the viewer. The challenge of seeing, imagining, and sketching or drawing in three dimensions is not just the reason of confusion in the study of stereochemistry. The other main cause is also the terminology that is used in it (Figure 2.1).

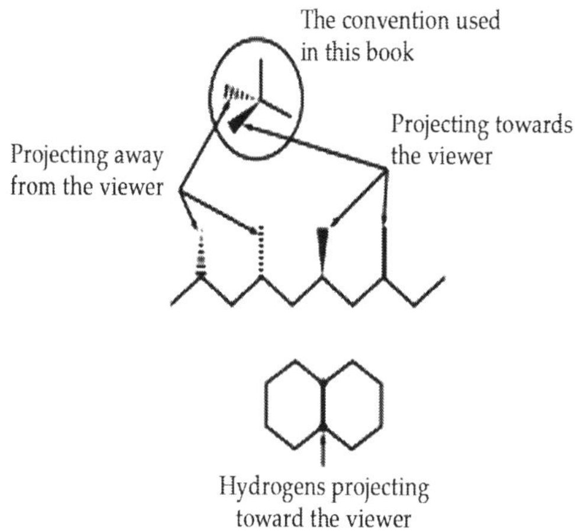

Figure 2.1. Three-dimension hydrogen.

Source: Image by uscibooks.com.

2.2. BASIC CONCEPTS AND TERMINOLOGY

It has been observed that in the current years, the stereochemical terminology has become crystal clear and definitive. We present here a discussion of the basics, not focused solely on carbon. We already know how stereoisomerism can give way to the formation of cis-and trans-2-butene. We also now understand how and why gauche as well as anti-butane are stereoisomers. Similarly, this contrasts with constitutional isomers, like 1-bromo-and 2-bromobutane of which the difference is seen in the connectivity of the atoms in the structural formula.

The constitution of a molecule is explained by the number as well as kinds of atoms and the connectivity of the atoms that makes up such molecule. It also includes bond multiplicity. Similarly, by now, you should have been aware of the distinction between configurational and conformational isomers.

Conformational isomers are the molecules that are interconvertible by rotations about single bonds. The conformation of a molecule concerns the aspects that are associated to rotations about single bonds. There exists some ambiguity to this difference, associated with the description of a "single" bond.

There are some related terms such as atropisomers. These are the stereoisomers that can be interconverted by rotation about single bonds. But for such stereoisomers, the barrier to rotation is large enough that the separation of stereoisomers is possible, and it do not interconvert readily at normal temperatures.

It has to be noted that the configurational isomer term is a historical term. In modern chemistry, it has no real value. Usually, it is used to include enantiomers and diastereomers as isomers. But the term stereochemical isomers are a better one.

The term configuration is still playing a very important role. Mislow describes configuration as "the relative order or position of the preparation of atoms in space which demonstrates a particular stereoisomer." One of the important related term in this is absolute configuration. This is basically concerned with the configuration of a structure settled upon stereochemical standard.

For instance, later in this chapter, emphasis will be put on d and L nomenclature system, where the preparation of atoms in space is associated with that of (+)-glyceraldehyde. If the preparation of atoms can be associated to (+)-glyceraldehyde, or some other standard in space in a molecule, then it means that it is a molecule's absolute configuration. It is generally seen that when two of the stereoisomers are non-superimposable mirror images of each other, they are considered as enantiomers.

In order to attain the mirror image of a molecule, just imagine a sheet of glass that is placed along with the molecule of interest. After this, pass every atom through the glass in the way that every single atom ends up the equal distance from the sheet of glass as it is in the original structure.

Diastereomers are the stereoisomers that are not enantiomers. Chiral object is any kind of object that is non-superimposable (noncongruent) with its mirror image. On the other hand, if the object is not chiral—that is, if the mirror image of the object is congruent with the original then it is achiral in nature.

2.2.1. Classic Terminology

There are various terms that are used in the context of stereochemistry. These terms are ingrained in the literature. And also, there are number of terms that one is familiar with since organic chemistry started. There are various research papers that described several terms and in addition to it, it also explained how all these terms can be misleading.

The uncertainty or confusion with regard to terminology happens with terms like "optically active" and "chiral center." Often, these terms mislead as much as they tell. Optically active is defined as the ability of a group of molecules to rotate the plane polarized light. A sample must have an excess of one enantiomer in order to be optically active.

Usually, optically active was used in the earlier literature as a synonym for chiral molecule and unfortunately this use of the term, sometimes, continues even in the present times. This use is not encouraged. The main issue is that there are several examples of molecules that have chiral sections, but is not optically active.

It has been observed that a racemic mixture, which is majorly a combination of enantiomers by 50:50 ratios is optically inactive. However, every single molecule in the sample is found to be chiral in nature. It is vital to differentiate between a sample, that is not optically active because it includes a racemic mixture, and a sample, that is not optically active because it includes achiral molecules.

Differentiating between the two samples was made difficult by the earlier terminology. In addition to it, it is very simple to imagine those molecules, even when these molecules are enantiomerically pure, that would not rotate the plane polarized light to any significant degree.

The extent to which the plane polarized light is rotated relies on the differences in the refractive indices with regards to right as well as left rotating polarized light as this light passes through the sample. There would be no measurable rotations if the enantiomers do not have different refractive indices.

Popular example can be hydrocarbon that have four different n-alkyl chains attached to it, with the length of the chain maybe 10, 11, 12, and 13 carbons; or one that is having four C_{10} chains, but terminating in–CH_3,–CH_2D,–CHD_2, and–CD_3.

In every single case that the molecule is chiral in nature, but the rotation of plane polarized light through the sample would be immensely small. Operationally, they are not optically active. It has been observed that a chiral molecule, even in an enantiomerically pure form, will show zero rotation at some wavelengths of light, as there is the movement from (+) rotation to (–) rotation in the optical rotatory dispersion (ORD) curve.

"Optically active" is not a defining description. The confusion happens with those terms that are meant to pay attention on the chirality at a specific point in a molecule. Chiral center or chiral carbon again, is particularly a carbon having four different ligands attached to it. The term "ligand" corresponds to any group that is attached to the chiral carbon, i.e., R, OH, H, Ar, etc.

Asymmetric carbon is the carbon that has four different ligands. The main issue with such kind of terms is that "asymmetric carbons" as well as "chiral centers/carbons" are in those molecules that are neither asymmetric in nature nor they are chiral. In addition to it, there are various molecules that can be in enantiomeric forms but has no "chiral center." Some of the classic examples consists of dimethyl allene and the twisted biphenyl.

2.3. TOPOLOGICAL AND SUPRAMOLECULAR STEREOCHEMISTRY

The preparation and characterization of molecules with novel topological features is one of the more interesting features of modern stereochemistry. It has been observed that a number of structures with novel topologies has been produced by Supramolecular chemistry. These products include catenates and rotaxanes.

There are some novel topological aspects of "Simple" molecules (that is not supramolecules) like knots or Möbius strips. The mathematical definition of topology, and the definition that is suitable or best suited to stereochemistry, emphasizes the studies of the aspects of geometrical objects that derive only from the patterns of their connectivity. Metric issues are linked with numerical values (like the bond lengths and bond angles)—are not important in topology.

One of the simplest and easiest way to see this is that the two-dimensional topology should be considered as the study of geometric figures. These figures have been drawn on a rubber sheet. The sheet can be stretched, and it can also be bent and flexed as much as one like to do without bringing any change in the topology of a figure that is drawn on the sheet.

Therefore, a triangle, a circle, and a square are topologically equivalent. This is because one can deform or bend one to the other. Topologically, all three (that is a circle, a triangle and a square) are just a closed loop. The same concept is applied in three dimensional structures of molecules. In addition to it, there are some additional requirements such as that one cannot break a line or permit any lines to cross, and also, a vertex cannot be destroyed.

In topography (that is map making), it matters how high the mountain is, but in the mathematical definition of topology we will use here, it does not (in fact, the mountain can be "stretched flat").

With some of the special exceptions that are discussed below, all the stereoisomers are, perhaps amazingly, topologically equivalent. If one is allowed to stretch the bonds and bend them at will, it is considered as a simple matter (Figure 2.2) to interconvert the enantiomers of 2-butanol. But this is done without crossing any bonds (mathematically simple but not chemically).

Same kind of distortions are possible with almost any of the molecule. This allows the stereoisomers to interconvert. This is consistent with the definition of stereoisomers as that kind of molecules that have the same connectivity (topologies) but these molecules have different arrangements of atoms in space.

Since, it has been observed that topology concerns about only the issues that are derived from the connectivity of the system, structures that have the same connectivity have the same topology. However, there are some stereoisomers with different topologies.

Topology is the study deals with graphs, those that concerns and that includes edges and vertices (the point where two or more than two edges meet). One is considering a chemical graph when considering the chemical topology. In this, the vertices are atoms and the edges are bonds.

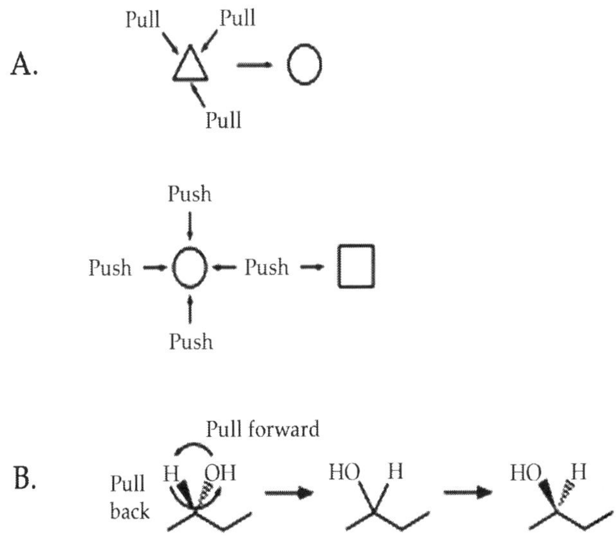

Figure 2.2. The inter-conversion of to pologically equivalent structures.

Source: Image by uscibooks.com.

A. Topologically, a circle, a triangle, and a square, are all just closed loops (as long as the "corners" of the triangle and square to be vertices are not considered).

B. The interconversion of the enantiomers of 2-butanol can be achieved by stretching and bending but this should be done without crossing any bonds. So, the two enantiomers are topologically equivalent in nature.

2.3.1. Loops and Knots

The simplest systems that can produce topological stereoisomers are cyclic structures. There is a circle and a classic trefoil knot that is shown in Figure 2.3. Both of the structures are shown in the figure are simply a single-closed loop (which is the definition of a knot, circle is the simplest knot or it can be called as the "unknot"). It is impossible to interconvert both of these structures by not crossing edges. Topologically, both of these structures are different.

There are some examples of topological stereoisomers such as the molecular realizations of the circle and the trefoil knot. It is logical to call them topological diastereomers since they are not non-congruent mirror images. A structure as simple as (CH_2) n could serve the purpose in order to make a chemical version of this situation.

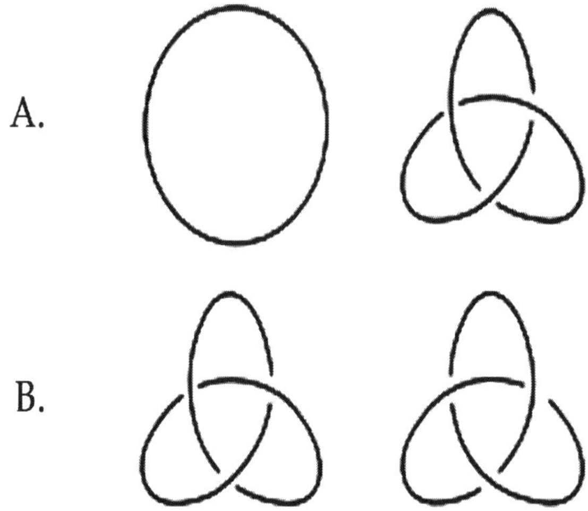

Figure 2.3. (A) Topological stereoisomers—a circle and a trefoil knot; (B) Enantiomorphs trefoil knots.

Source: Image by uscibooks.com.

How are these stereoisomers different from that of conventional diastereomers? It has to be noted that the circle and the knot can be infinitely deformed that is, it can be twisted, curved, stretched, and compressed but inter-conversion can never take place (as long as no bonds are crossed).

By the process of deformation, conventional isomers can be inter-converted as in the case of 2-butanol in Figure 2.3. Conventional stereoisomerism is reliant on the exact location of the atoms in space. This will result to the terms geometric isomerism or Euclidian isomerism. The atoms are moved all around, and isomerism can be retained with topological stereoisomers.

2.3.2. Topological Chirality

If there are possibilities of topological diastereomers, then there must be a possibility of topological enantiomers. This means that there is topological chirality. The trefoil knot is a simple example of this. In case of molecules, the term enantiomer is reserved and enantiomorphs is applied to geometrical objects. There are many researches who are curious to find out the probability of just deforming one structure into the other by stretching and pulling. If this is possible, the two forms would be topologically equivalent and therefore, these will not be enantiomers. The trefoil knot would be topologically achiral in nature.

Also, there exist no general way that can be used to prove that a knot is chiral in nature. It can be proved as achiral by only discovering that single way to draw the knot (that is known as a presentation) that is itself achiral in nature.

Though, if someone fails in finding an achiral presentation, it does not prove that the knot is chiral in nature; The possibility is that the achiral presentation cannot be found. However, the structure is indeed chiral in the case of the trefoil knot.

2.3.3. Non Planar Graphs

There are several numbers of topological issue that has fascinated many of the chemists. For most of the organic molecules, a two-dimensional representation can be drawn. It has no bonds that crosses each other. This is known as a planar graph. If the connectivity of a system cannot be represented without some crossing lines, there is a non-planar graph. It might seem astonishing, but most of the molecules have planar graphs. It should be kept in mind that while conducting topology, one can stretch the bonds and also bends them as per their requirements.

Graph theory is defined as a mature branch of mathematics. It has been established by the graph theorists that all non-planar graphs will conform to one of two prototypes, which is known as K_5 and $K_{3,3}$ in graph theory terminology.

There are five vertices in K_5. It is maximally connected. In K_5, each vertex is connected to each other. There are two sets of three vertices in $K_{3,3}$. In this, each vertex of one set is linked to each vertex of the other set. The fact that $K_{3,3}$ is non-planar in nature explains the architectural conundrum of three houses and three utilities. It is not possible to have three houses that are linked to three utilities (like electric, water, and phone) without even a single instance of lines crossing or overlapping of each other. The molecular versions of these non-planar graph are represented Figure 2.4.

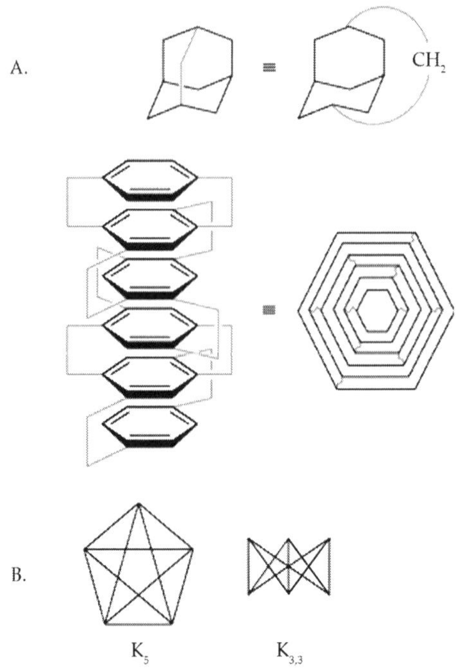

Figure 2.4. (A) Examples of how most of the chemical structures can be characterized as planar graphs, and (B) K_5 and $K_{3,3}$ nonplanar graphs.

Source: Image by uscibooks.com.

2.3.4. Achievements in Topological and Supramolecular Stereochemistry

Many recent attempts have formed the chemical structures that have effectively realized many of interesting as well as novel topologies. A landmark was definitely the process of synthesis of a trefoil knot by the use of Sauvage Cu+/phenanthroline templating strategy.

This structure that is nonplanar and chiral in nature is considered as a standard for the field. In addition to it, there were some other more complex knots that was also prepared by the use of this strategy. The team of Vogtle and other scientists have defined an "all organic" approach to amide-containing trefoil knots.

Also, they were able to separate the two enantiomeric knots by the use of chiral chromatography. The process of synthesis and characterization of a "Möbius strip" molecule (Figure 2.5) was the other seminal advancement.

A Möbius strip is believed to be a closed ribbon that has a twist. In addition to it, it also has long fascinated mathematicians as well as the general public for a long period of time.

In the 1950s, the concept that is behind the Möbius strategy for the preparation of novel topologies was enunciated. Until the 1980s, it was not chemically realized. Walba developed a clever strategy that is based on tetra hydroxymethyl ethylene (THYME) ethers.

There are possibilities that ring closure could proceed in the presence or twist or in the absence of a twist. The two are formed in almost equal amounts when the reaction is carried out. The most significant design feature was that the "rungs" of the ladder system were olefins. This could be selectively split by the process of ozonolysis. The gap between the untwisted product formed two small sized rings, but the gap of the Möbius product produces a single, larger macro-cycle. Therefore, distinguishing the two topological stereoisomers.

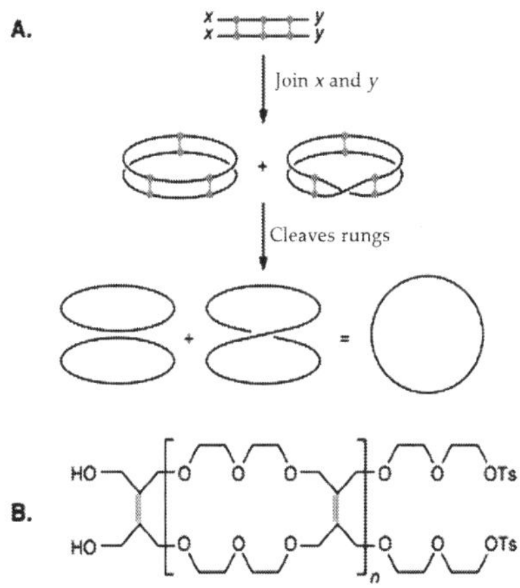

Figure 2.5. A. The synthetic strategy which is majorly used for the formation of a molecular Möbius strip, and the results of rung cleavage and B. THYME polyether that can come closer to each other in order to form a + Möbius strip.

Source: Image by uscibooks.com.

It has to be noted that the three-rung Möbius ladder compound is a molecular realization of an interesting topology even in the absence of the twist. It is considered as a simple example of a nonplanar graph which have $K_{3,3}$ topology.

2.4. TOPOLOGICAL JUNCTIVE/DISJUNCTIVE PROCESSES

By the use of the notations for fundamental junctive simplexes that are mentioned above, each of the hypothetical transformations that is shown in Figure 2.6., viz. 106+107-»108,109+110-»lll, 112a+112b—113, and 114, 115, consists of two separates $(l, l)_j$ processes. It could be defined as a (l, l) j $(l, l)_j$ or $(l, l)_j^2$ process.

Though, the 2 junctive N atoms, in each of 106 and 109, carries a solo topological bi-junctive moiety; in a similar way, both junctive boron sites in each of 107, and 110.

For the concept of 112a (112b), both the B and the N atoms in the similar molecule also signify a solo topological bi-junctive simplex. It is clearly observable that these conversions signify a shared topological $(2_t,2_t)_j$ process (where 2_t labels a topological diplex or topological bi-junctive simplex). These examples exhibit influentially that the notion of junctions can also be applied to be intricate molecular systems with multi-junctive sites.

The said extension provides the groundwork for topological junctive, with the basic units being labelled as topological simplexes. A topological simplex refers to a meagre group of junctive atoms (in a portion of molecule or in entire molecule) that take part in a junctive process embracing another small group of junctive atoms in the distinct or identical molecule. The atomicity of a topological simplex (a_t) can be defined as the total of junctive atoms in the small group defined above.

Thus, every molecule 106, 107, 109, 110, and 112a (112b) holds a topological bi-junctive simplex; molecule 114 retains two such simplexes. These examples exhibit that (a) a single molecule may carry one or beyond that topological simplex (es), and (b) an existing or present topological simplex should be integrated into one and the identical molecule for instance, portion of a topological simplex cannot be in distinctive molecules/ molecular fragments. The associative engagement between two and three topological simplexes (vide infra) resulting in forming simple topological junctive processes; these are signified as (m_t, n_t)j, $(m_t, n_t, p_t)_j$ for termolecular and bimolecular associations, correspondingly (m_t, n_t, p_t refers to atomicity of the engaging topological simplexes i.e., m, n, p = l, 2; subscripted suffixes j and d, mean disjunctive and junctive, correspondingly; where t denotes to topological). The corresponding reverse (disjunctive) processes can be labelled as (m_t, n_t)d, $(m_t, n_t, p_t)_d$. Some depiction of such processes is demonstrated in Figure 2.6, along with the abbreviated designations.

Some of the cases of binary associations are demonstrated by biplex/atom $(l,2_t)$–116-»[117], and biplex/biplex $(2_t,2_t)$–118-»[119]. In a similar way ternary associations atom/biplex/atom $(l, l,2_t)$, biplex/atom/biplex $(l,2_t,2_t)$, and biplex/biplex/biplex $(2_t,2_t,2_t)$ are signified by 120-»[121], 122-»[123], and 124-»[125], correspondingly. Some molecular examples of topological junctive processes have been demonstrated in Figure 2.6. The meagre amount of p-hairpin of 127 is produced by a $(2_t,2_t)$ junctive process of 126; while on the other hand, the complexation of Li^+ by 15-crown–5, is a $(1,5,)_j(l, l)_d$ process. The pairing of adenine-thymine (131+132—133)11 is a $(2_t,2_t)_j$ process, while 134+135-»13612 is a $(3_t,3_t)_j$ process. Rebek's dimerization 137a+137b-»138[13] is a $(2_t,2_t)_j^2$ process, whereas the grouping of diamine dioxide 140 with respect to hydrogen peroxide 14 is a $(l,2_t, l)_j$ process. The accumulation of 142+143a+143b[10] is a $(2_t,2_t)_j(3_t,3_t)_j$ process, and the tetramerization of 145a-d-»14615 is a $(2_t,2_t)_j^4$ process.

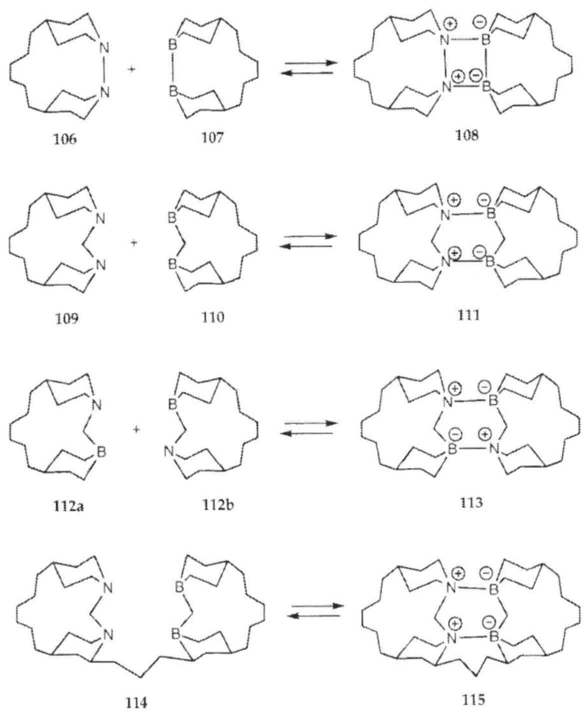

Figure 2.6. Hypothetical examples of intermolecular and intramolecular examples of topologically junctive processes.

Source: Image by Elsevier.

2.5. LIGOGENIC/LIGOLYTIC PROCESSES IN ORGANIC CHEMISTRY

Ligogenic/lipolytic processes basically comprises of a subgroup that involves disjunctive/junctive alterations; they carry separate π and/or σ+ π bond cleavage and/or creation; associations through H-bonding, and complexation through π-bonding are particularly omitted. Thus, it can be said that every ligogenic process is essentially junctive; Although, every junctive process is not certainly ligogenic. In a similar way, each lipolytic process is inevitably disjunctive; every disjunctive process is not certainly lipolytic (vide infra).

2.5.1. Ligogenic/Ligolytic Processes

A ligogenic process can be defined as a one in which separate π and/or σ+ π(but not just π) bonding occurs in-between two unbonded atoms.

In case of canonical structures, it is possible to use localized bonding models. In a ligogenic process, the order of bond between two atoms that are reactive in nature starts at 0 and intensifies to integral values which is higher than 0 for instance 1, 2, 3.... An aligogenic process can be defined as one where there is likelihood of occurrence of junctivity, but without creation of any π bond e.g., in the creation of π-complexes or H-bonded species.

A lipolytic process refers to the particular opposite of a ligogenic process {vide infra). In a lipolytic process, for two atoms that are connected with each other, one can be started at any integral value of a bond order which is greater than 0 and ends up at 0. In Figure 2.7, it can be depicted that the alteration in bond order that follows each lipolytic and ligogenic process. In Figure 2.7, complexation of guest-host l+2-*3, π complexation 4+5-»6, dimerization through H bonds 7a+7b-»8 all are junctive conversions; while it is important to note that they are not ligogenic alterations because in none of them is there construction of distinct π and/or σ+ π bonds. In contrast, the overall conversions 9+10-»ll, 12a+12b-»13, 14+15-»16, 17+18-»19, and 20+21-»22 are ligogenic junctive processes because of the fact that every one of the them embraces creation of new π and/or σ+ π bonds.

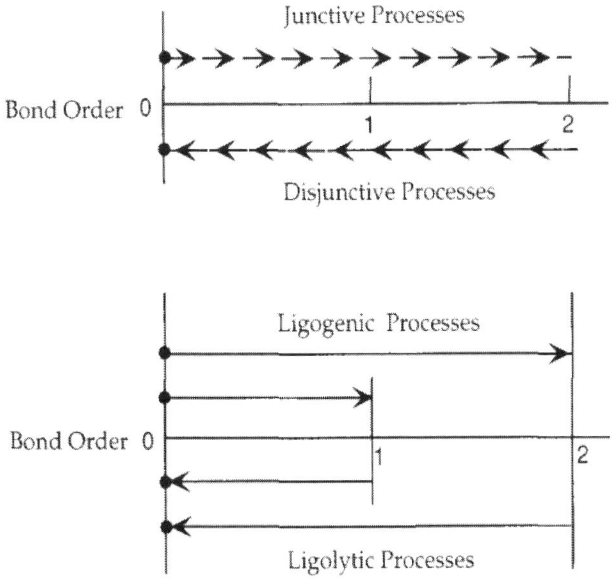

Figure 2.7. Changes in bond order in junctive/disjunctive processes vs. ligogenic/lipolytic processes.

Source: Image by Elsevier.

2.6. MORPHO SELECTIVITY

In organic reactions where several products or molecules are formed by a chemical reaction, one of these products are classified as the major product while the rest were minor products. Thus, the tendency of a chemical reaction to a major and the minor products depends on several conditions met during the reaction process, indicating that the ability of the process to distinguish such major product over the possible products of the reaction refers to the selectivity of the chemical reaction. Thus, selectivity refers to the expression of one of the basic characteristics of chemical transformations over the other characteristics. It is a complex concept conveyed through an assembly of various terms that portrays opposing facets of selectivity. The diversity or comprehensive range of terms of selectivity in the chemical literature is an outcome of the multiple requirements to denote; (a) the rate or degree of consumption of one of two (or more) reactants, (b) the rate or degree of development of one of two (or more) probable products (with reference to either constitution or stereochemistry), or (c) the inclination towards reaction site(s) (e.g., bond, atom, functional group, molecular face) in a molecule. As the criteria used in this is very complicated, it is doomed, but not astonishing, that the terms are sometimes used inaccurately or imprecisely. This section is mainly concerned with morpho selectivity, and tries to establish a relationship between the above terms.

2.6.1. Morpholytic Selectivity vs. Morphogenic Selectivity

For a specific molecular transformation, the major area of interest is either on the consumption of a reactant S ("morpholytic process," Gr. morphe, shape, form; unbinding, lysis) or, the formation of a product P ("morphogenic process," Gr. morphe, shape, form; L. generation, genesis)

Morpho selectivity is the selective use of reactant S1 over S2 (morpholy to selectivity), whereas that exemplifying the chosen generation of product P1 over P2 is termed as morphogenic selectivity (morphogeno selectivity). The corresponding processes can be referred as morpholy to selective and morphogeno selective, correspondingly. The establish relationship between selective morphogenesis and selective morpholysis is represented in Figure 2.8.

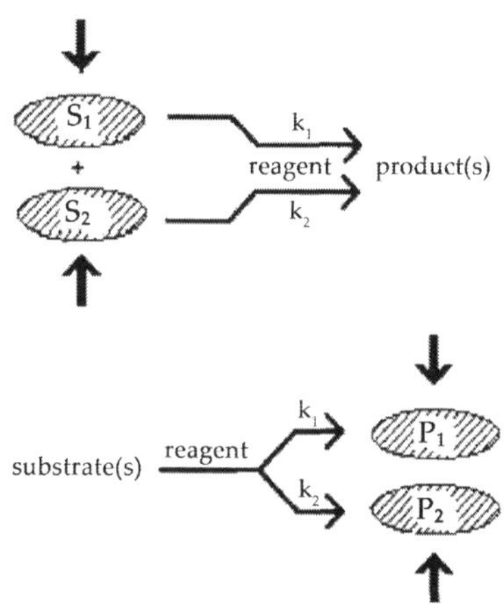

Figure 2.8. Idealized representation of competitive morpholytic processes (Sj vs. S2) and morphogenic processes (Pj vs. P2).

Source: Image by Elsevier.

2.6.1.1. Morpholytic Selectivity

It is important to note that a morpholytic process can be selective, as elective or nonselective. A reaction considering the substances S1 and S2 as reactants, is conceived to implicate selective morpholysis, if S1 is expended at a more rapid rate (or at a slow rate) compared with S2 (S1*S2). In case of nonselective morpholysis, both S1 and S2 are consumed at an anticipated or accidentally at a similar rate. While, on the other hand, an as elective morpholytic process refers to the one in which selective intake of reactant (s) is operationally unnoticeable, because of the existence of only one reactant substrate S (i.e., S1=S2=S), irrespective of whether the reaction resulted in producing one or more products.

One reactant substrate configuring into more than one product may be labeled as elective divergent morpholysis. The specific category of morpholytic selectivity can be described by the morphic relationship between S1 and S2 {vide infra).

In case both S1 and S2 are stereometric in relation to each other, then the process is depicted by stereomorpholytic selectivity. More precisely, if S1 and S2 are, diastereomeric or enantiomeric, the process is believed to display diastereomorpholytic selectivity orenantiomorpholytic selectivity, correspondingly. It is worth noticing that use of homomorphic molecules (homomeric substances) is a selective in nature; therefore, there is no such thing like selective homomorpholysis. While, on the other side, if S1 and S2 are nonstereomeric in relation to each other, the activity is termed by nonstereomorpholytic selectivity.

If S1 and S2 are nonequimeric or astereomeric with each other, the practice is said to portray nonequimorpholytic selectivity or astereomorpholytic selectivity, correspondingly. For selectivity in complexative or junctive processes, the corresponding terms are enantiomorph junctive, stereomorphojunctive, diastereomorphojunctive, astereomorphojunctive, nonstereomorphojunctive, and nonequimorphojunctive. It is important to note that there is significant difference between stereomorpholytic nonselectivity and nonstereomorpholytic selectivity.

The latter term denotes to selectivity for two reactants which are nonstereomeric with each other, whereas the former term denotes the non-existence of selectivity in the process where reactants are stereometric with each other. In next few sections, all these terms are discussed in detail with few specific examples from the chemical literature.

2.6.1.2. Morphogenic Selectivity

A given morphogenic process is also subclassified as nonselective, selective, or as elective. A reaction with two or more products such as (P1, P2. Pn.) is said to display selective morphogenesis, if product P1 is constructed in liking to product P2 (P1*P2).

Here, both P1 and P2 are discrete main products through two separate (nonchronometric) pathways beginning from either a solo or 2 substrates. In contrast, as far as nonselective morphogenesis is concerned, both products P1 and P2 are attained in accidentally or anticipated similar rates. To distinguish, in an as elective morphogenic process, it is almost unascertainable to have the selective formation of product (s) primarily because of the fact that only one product can be formed P (i.e., P1=P2=P), in reactions that may initiate from one (or more) starting material(s).

The case concerning two reactant substrates transforming to a single product may be labeled convergent morphogenesis. Where there is occurrence of morphogenic selectivity, the category of selectivity is in context to the morphic relationship between products P1 and P2 (vide infra). If P1 and P2 are stereometric with each other, then the process is outlined by stereomorphogenic selectivity.

More precisely, if P1 and P2 are diastereomeric and enantiomeric with each other, the process is said to display diastereomorphogenic selectivity orenantiomorphogenic selectivity, correspondingly. While, on the other side, for reactions ruling selectively to nonstereomeric products, the processes are depicted by nonstereomorphogenic selectivity. Specifically, if P1 and P2 are nonequimeric or astereomeric with each other, the process is said to demonstrate nonequimorphogenic selectivity or astereomorphogenic selectivity correspondingly.

It is further worth considering that selective stereomorphogenesis is identical with stereoselective synthesis. More precisely, diastereo morphogenesis, and selective enantiomorph genesis are identical with diastereo selective syntheses and enantio selective, correspondingly. It is operationally uncertain to form homomeric substances through competing pathway; hence, there is no such thing as selective homomorphogenesis.

Similarly, the difference between stereo morphogenic nonselectivity and nonstereomorphogenic selectivity can be clearly observed (vide infra). The latter term denotes to the selectivity in the yield of two products that are nonstereomeric with each other; the former term denotes to the nonexistence of selectivity in a process where products are stereometric with each other.

2.7. CHIRALITY AND SPATIAL ARRANGEMENT OF ATOMS

Organic compounds and molecules formed along with a chain of carbon atom (generally recognized as carbon backbone), play very pivotal role in the life of chemistry. Their value or importance are usually from the energy they carry, particularly in an arrangement of potential energy between atomic molecules.

Because this potential force can be significantly affected due to variation in atomic placement, it is essential to gain insights into the concept of an isomer-a molecule sharing same atomic make up as another but varying in structural arrangement. In this section, the focus is on the specific isomers referred as stereoisomers and its chirality (Figure 2.9).

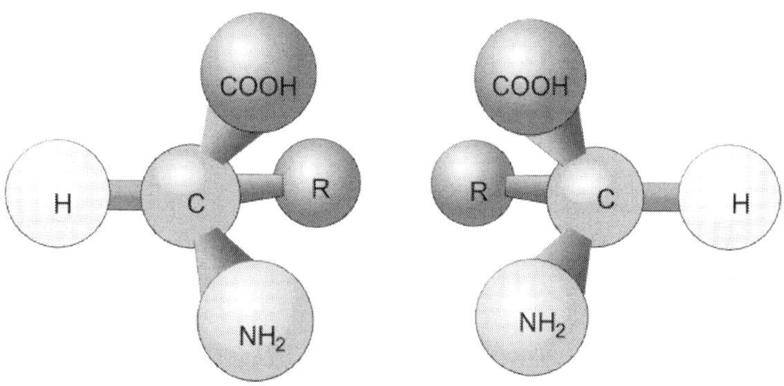

Figure 2.9. Two enantiomers of a tetrahedral complex.

Source: chem.libretexts.org.

The concepts of stereoisomerism and chirality is of high importance in the modern-day organic chemistry, as these helps to build an understanding of theoretical and physical reasons behind the structures and development of numerous organic molecules, with an underlying basis on the theories and concepts on the energy integrated in these vital chemicals.

For instance, to widely recognize constitutional isomerism, which forms isotopic compounds lucidly by distinct atomic connectivity, stereoisomerism usually entails equal atomic connections and arrangement of building blocks as well as carrying the identical numbers of atoms or elements.

Then what makes stereoisomers so unique? To explanation this, there is need to recall and imagine in not just the 2D images, but also 3D configurations. It is primarily because stereoisomers are isomers where their atoms are distinct from each other in terms of their spatial arrangement.

2.7.1. Spatial Arrangement

Spatial arrangement of atoms is concerned with how distinct molecules and atomic particles are allocated in and around the space of organic compound, viz., its carbon chain.

In this manner, it is right to assume that spatial arrangement of an organic molecule varies from each other if an atom is transferred in any 3D direction even by a meagre amount. This widens the likelihood of different molecules, each with their eccentric placement of atoms in 3D space.

2.7.2. Stereoisomers

Stereoisomers as described above and in the previous sections of this book, entails variety of isomers within itself, each having unique characteristics that further express each other as distinct chemical entities having different properties. Enantiomer are the mirror-image stereoisomers which are mentioned earlier and will be elucidated comprehensively in this section, in addition, this section will also include diastereomers, with its unique properties being highlighted.

2.7.3. Enantiomers

The stereoisomers that are mirror-images of each other but are non-super imposable are called enantiomers. One of the perfect examples are provided below; it is worth noticing that the gray plane in the middle demonstrates the mirror plane (Figure 2.10).

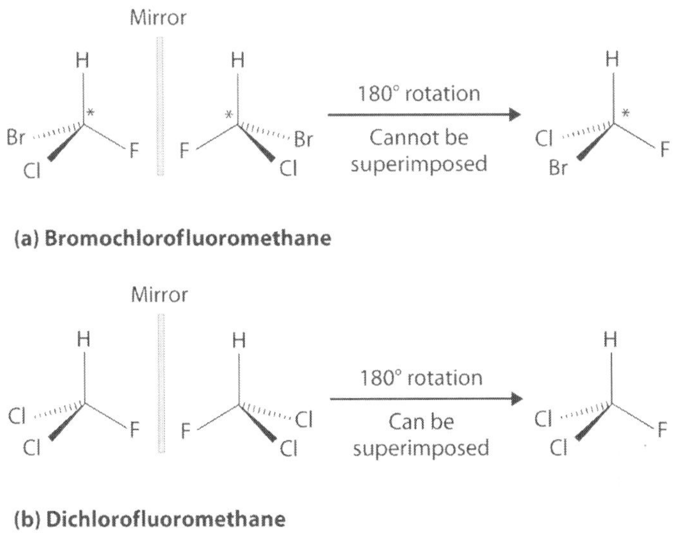

Figure 2.10. *Comparison of chiral and achiral molecules.*

Source: chem.libretexts.org.

A. Bromochlorofluoromethane is a chiral molecule whose stereo center is appointed with an asterisk. Revolving or rotating its mirror image does not necessarily produce the original structure. To superimpose the mirror images, it is essential that bonds should be reformed and be broken.

B. In contrast, it is possible for dichlorofluoromethane and its mirror image to be superimposable at a specific angle of rotation.

It is important to note for enantiomers that even the left molecule is flipped over to the right, the atomic arrangement in space will not be identical. This is similar to the left hand and right hand overlapped with each other., and that no matter how you try, the left hand will fit with the structure of right hand in any angle you rotate the other (you are not allowed to lift any of the hand and face it with each other, because that is the mirror image). Because of this, enantiomers are described to exhibit "handedness" and since enantiomers are chiral molecules, handedness, and chirality are associated with each other.

2.7.4. Chirality

Chirality mainly refers to the non-superimposable mirror images of the molecule/s. Therefore, describing an organic compound or molecule chiral, indicates that aside from itself, it assumes additional forms or structure. Whether a molecule is chiral or achiral, therefore, primarily relies on a variety of overlapping conditions. In Figure 2.11, an example of two molecules portrayed, achiral, and chiral, respectively.

It is important to take into consideration the unique characteristics of the achiral molecule: it mainly carries two atoms of same element. In both reality and theory, if one wants to form a plane that runs through the other two atoms, they can be able to produce what is known as bisecting plane: Take note of the images on both sides of the plane in (a), which is as same as the other (Figure 2.11).

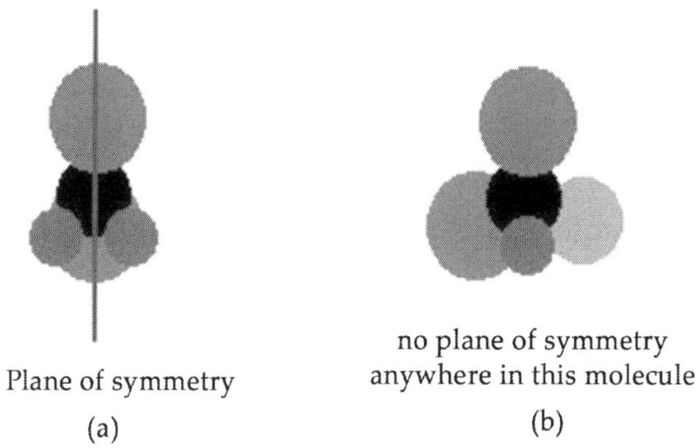

Figure 2.11. (a) Achiral molecule usually yields mirror images of the other half when placed with a plane of symmetry. (b) While chiral molecules does not have a plane of symmetry.

A chiral molecule has four different groups attached to it, so it is nonsense to use a plane of symmetry, since there will be basically no symmetry to be achieved anywhere the imaginary plane is placed. Thus, this generalizes the fact that chiral molecules do not have a plane of symmetry.

On the other hand, achiral molecules with two similar groups, has a plane of symmetry provided each of the similar group is on each of the sides of the plane just like what is illustrated in Figure 2.11(a). This makes the plane of symmetry highly associated with achiral molecules.

To generalize, the presence of the four different groups attached to the central Carbon atom, or to the carbon backbone (when dealing with larger molecules), makes chiral molecules asymmetric through a cutting plane. Conversely, the plane of symmetry is more defined when the two similar substituent's of the central carbon atom or carbon backbone of an achiral molecule lies on each sides of the plane.

2.7.5. Compounds with Multiple Chiral Centers

Can molecules carry more than one stereo center? This is one of the basic questions encountered in discussions involving chirality, and the answer is "yes." Classic example is a four-carbon sugar called D-erythrose.

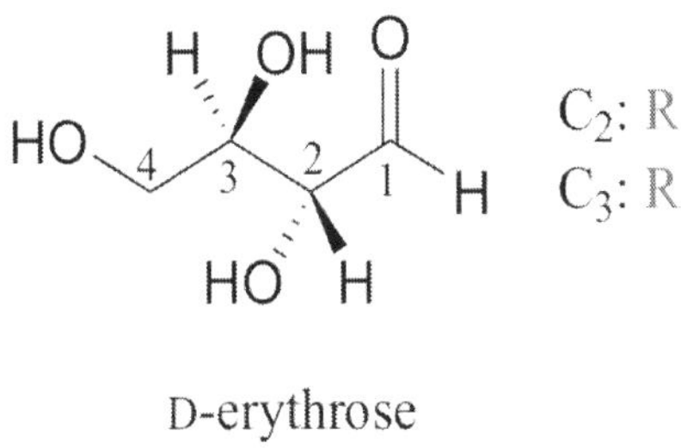

Figure 2.12. Compounds with multiple chiral centers.

Source: chem.libretexts.org

An important note on sugar nomenclature: biochemists generally make use of a special system in order to refer to the stereochemistry of sugar molecules, using names of historical era in extension to the designators '*L*' and '*D*.' This system can be learned in detail if one takes the biochemistry class. In this section the *D/L* designations will be used primarily to different sugars

As one can clearly see, *D*-erythrose is a chiral molecule: C_3 and C_2 both are stereo centers that entails the *R* configuration. Moreover, it necessitates the need to make a model in a way to convince oneself it is not possible to find a plane of symmetry through the molecule, irrespective of its conformation. It is further important to ask does D-erythrose have an enantiomer. Of course, it does–if it is a chiral molecule, then it definitely has one. The enantiomer of erythrose refers to its mirror image and is mainly known as L-erythrose (once again, there is need to make use of models in a way to convince oneself that these mirror images of erythrose are not superimposable) (Figure 2.13).

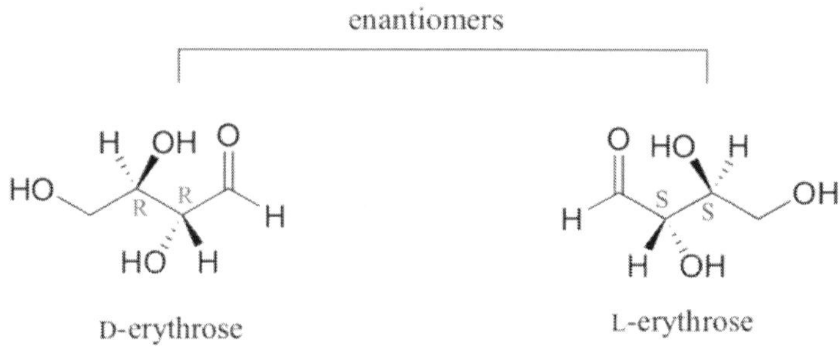

Figure 2.13. Relationship between D and L enantiomers.

Source: chem.libretexts.org

It is important to note that both chiral centers in L-erythrose carries the *S* configuration. In a set of enantiomers, the entire chiral centers carry the R configuration.

One important question that can be asked in this is that what happens in case one draws a stereoisomer of erythrose in which the configuration is *R* at C_3 and *S* at C_2? This stereoisomer that is a sugar referred as D-threose, is *not* a mirror image of erythrose. D-threose is often known as a diastereomer of both L-erythrose and D-erythrose.

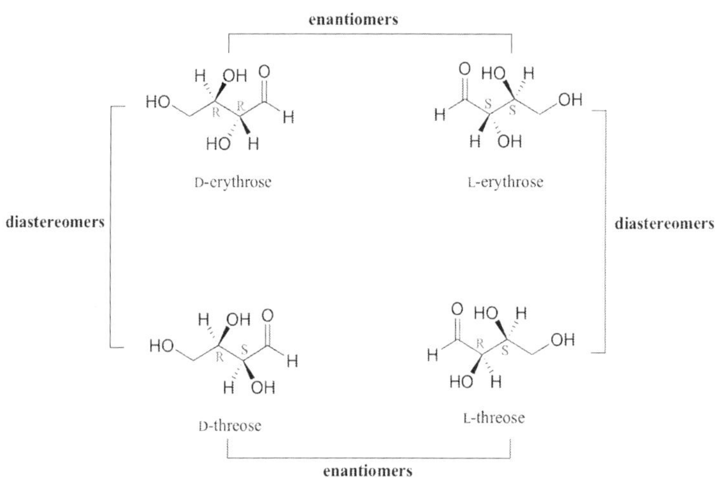

Figure 2.14. Enantiomers and diastereomers.

Source: chem.libretexts.org

The meaning of diastereomers can be easily understood: if there are two molecules which are stereoisomers (similar connectivity, identical molecular formula, distinct arrangement of atoms in space) but are *not* enantiomers, then by default, they diastereomers. In practical terms, it can be considered that at least 1-if not entire-chiral centers are inverse in relationship in a pair of diastereomers. By definition, the two molecules which are diastereomers cannot be considered mirror images of each other (Figure 2.14).

L-threose, the enantiomer of D-threose, has the *R* configuration at C_2 and the *S* configuration at C_3. L-threose is a diastereomer of both erythrose enantiomers.

In general, a structure having *n* stereocenters will likely to entails 2^n distinct stereoisomers. For instance, let's take into consideration the glucose molecule in its open chain form (recall that majority of the sugar molecules possess the ability to remain in existence either a cyclic form or an open-chain). It is important to note that there are 2 enantiomers of glucose, referred as L-glucose and D-glucose. The D-enantiomer is known as common sugar that human bodies utilize for energy. It has $n = 4$ stereocenters, therefore there are $2^n = 2^4 = 16$ possible stereoisomers (comprising D-glucose itself).

2.8. STEREO ISOGRAMS OF OCTAHEDRAL COMPLEXES

Since Werner introduced coordination theory, the stereochemistry of octahedral complexes has received great interest in inorganic stereochemistry. Among the various concepts of inorganic stereochemistry, in general, problems in relation with symmetrical and enumerational properties have consistently attracted attentions of theoretical and in organic chemists.

As for the difficulties in enumerational properties, Polya's theorem and computer enumeration have been widely used to compute to count octahedral derivatives in equivalent ways. The approach for such enumerational difficulties has been re-identified by initiating the unit-subdued-cycle-index (USCI) approach, where the novel development approach for organic compounds has been effectively affected to the enumeration of octahedral complexes.

While on the other side, symmetry problems of octahedral complexes have only been limited to simple applications of the conventional methodology of organic stereochemistry. For instance, the section title "Describing complete configuration—differentiating between enantiomers" of the IR–9.3.4 of the international union of pure and applied chemistry (IUPAC) recommendations version 2005 suggests that the *C/A*-system for defining absolute configurations of octahedral centers aims at signifying to "chirality."

This is almost related to the state of the art for the *R/S*-stereo descriptors of the Cahn-Ingold-Prelog (CIP) system for compounds which are organic in nature, where the widely used term "chirality center" is now related to the term "stereogenic," and was embraced as its theoretical foundation in place of "chiral." In addition, the "prochirality" concept has been mainly used for treatment of organic compounds, which is mostly applied to the debate on prochirality of coordination compounds. Thus, it can be said that the original construction for organic compounds has been intuitively assumed to be a geometrical concept (cf. the term "prochirality" of the IUPAC Basic Terminology of Stereochemistry) and has been functional to coordination compounds. Conversely, it is widely believed that the tangled situations between "stereogenicity" and "chirality" as well as between "prostereogenicity" and "prochirality" (or "prostereoisomerism") in organic stereochemistry have been unintentionally carried into inorganic stereochemistry.

The complex scenario in organic stereochemistry have been circumvented by stereo isograms, which was projected to gain an insight about stereochemical phenomena. Thus, a quadruplet of *RS*-stereoisomeric (pro) molecules form a stereoisogram, which entails two groups of enantiomers, two groups of *RS*-diastereomers, and two groups of Hola timers. Such sort of stereo isograms have been mainly categorized into five types, where the presence of the five types has been proven by considering *RS*-stereoisomeric groups.

The stereo isograms concept has been preliminary applied to tetrahedral skeletons, where the key goals are organic compounds. A stereoisogram possess the ability of specifying chirality and *RS*-stereogenicity self-reliantly by initiating *RS*-diastereomeric, enantiomeric, and holantimeric relationships. Thereby, *R/S*-stereodescriptors under the system of the CIP are elucidated to be resolute by *RS*-stereogenicity and not by chirality. Then, the *R/S*-stereo descriptors on the basis of *RS*-stereo genicity are associated with the chirality through chirality.

In addition, well established oblivious confusion on the terms "stereogenicity" and "prochirality" in the conventional stereochemistry has been exposed and entirely developed by ways of stereo isograms. Substitution criteria for computing the pro-*RS*-stereogenicity and prochirality has been settled based on stereo isograms, where pro-*R*/pro-*S*-descriptors are elucidated to be listed by relationship of *RS*-diastereotopic, but not by stereohetero-topic relationships.

A forward-thinking concept of correlation diagrams of stereo isograms has been projected to the stereoisomers of cyclobutene derivatives as well as uninuclear and binuclear promolecules. More complex conducts have been critical to depict ethylene derivatives, allene derivatives, where *RS*-stereoisomeric groups, point groups, is skeletal groups, stereoisomeric groups are introduced.

While on the other hand, for inorganic complexes, the notion of stereo isograms has been applied to square planar complexes, which are scrutinized by means of *RS*-stereoisomeric groups, point groups, and stereoisomeric groups. Stereo isograms of octahedral complexes, on the other side, have not been learned because of that fact that it is highly complicated due to their high symmetries.

Because octahedral complexes play very crucial role in inorganic chemistry as explained above, the next goal is to gain understanding of octahedral complexes, which will be unraveled by ways of stereo isograms in the existing paper, so as to determine the adaptability of stereo isograms in inorganic stereochemistry.

2.9. CONCLUSION

In the chapter, the focus is to try to resolve all the confusion that can be generated with the terms by giving detailed description of various concepts. Some was eliminated by the introduction of stereogenic center (orin other words, stereocenter) as an organizing principle in stereochemistry.

If the interchange of two ligands attached to a central atom in a molecule gives way to the generation of a new stereoisomer, that central atom is a stereogenic center. However, not all interchanges give a new stereoisomer, but if one does, then the center can be conceived as stereogenic. Therefore, the center "generates" stereochemistry. A non-stereogenic center refers to one that exchanges any pair of ligands which does not generate a stereoisomer.

It is also important to note that the term "stereogenic center" is, in a sense, much broader in application compared with the term "chiral center." It implies nothing about the molecule being chiral, instead, it tells us that stereoisomerism is possible. In addition, it is also worth noticing that for more complex geometries, such as hexaco-ordinate or pentaco-ordinate atoms, there is no need for all the ligands to be in-equivalent in a way to have a stereogenic center.

REFERENCES

1. (2007). Module 8 Stereochemistry Lecture 20 Stereochemistry I. [ebook] Available at: https://nptel.ac.in/content/storage2/courses/104103071/pdf/mod8.pdf (accessed on 18 August 2020).
2. (2011). Stereochemistry: Chiral Center Representation. [ebook] Available at: http://www.siue.edu/~tpatric/ster.pdf (accessed on 18 August 2020).
3. (2015). Basic Concepts in Organic Chemistry and Stereochemistry. [ebook] Available at: https://doubt.dubbtr.com/assets/uploads/files/1537091883118-1537086550387-7basic-concepts-in-organic-chemistry-and-stereochemistry.pdf (accessed on 18 August 2020).
4. Chemistry LibreTexts, (2019). 3.12: Prochirality. [online] Available at: https://chem.libretexts.org/Bookshelves/Organic_Chemistry/Book%3A_Organic_Chemistry_with_a_Biological_Emphasis_v2.0_(Soderberg)/03%3A_Conformations_and_Stereochemistry/3.12%3A_Prochirality (accessed on 18 August 2020).
5. Chemistry LibreTexts, (2019). 6.1: Chirality. [online] Available at: https://chem.libretexts.org/Courses/Sacramento_City_College/SCC%3A_Chem_420_-_Organic_Chemistry_I/Text/06%3A_Stereochemistry_at_Tetrahedral_Centers/6.01%3A_Chirality (accessed on 18 August 2020).
6. Chemistry LibreTexts, (2019). Chirality and Stereoisomers. [online] Available at: https://chem.libretexts.org/Bookshelves/Organic_Chemistry/Supplemental_Modules_(Organic_Chemistry)/Chirality/Chirality_and_Stereoisomers (accessed on 18 August 2020).
7. Chemistry LibreTexts, (2020). 25.7: Chirality in Organic Chemistry. [online] Available at: https://chem.libretexts.org/Bookshelves/General_Chemistry/Map%3A_Chemistry_-_The_Central_Science_(Brown_et_al.)/25%3A_Chemistry_of_Life%3A_Organic_and_Biological_Chemistry/25.07%3A_Chirality_in_Organic_Chemistry (accessed on 18 August 2020).
8. Chemistry LibreTexts, (2020). 5.1: Enantiomers and the Tetrahedral Carbon. [online] Available at: https://chem.libretexts.org/Bookshelves/Organic_Chemistry/Map%3A_Organic_Chemistry_(McMurry)/05%3A_Stereochemistry_at_Tetrahedral_Centers/5.01%3A_Enantiomers_and_the_Tetrahedral_Carbon (accessed on 18 August 2020).

9. Fujita, S., (2012). Stereoisograms of Octahedral Complexes. I. Chirality and RS-Stereogenicity. [online] Match.pmf.kg.ac.rs. Available at: http://match.pmf.kg.ac.rs/electronic_versions/Match71/n3/match71n3_511–536.pdf (accessed on 18 August 2020).
10. Kaloustian, M., (2004). Concepts and Terminology in Organic Stereochemistry. Amsterdam: Elsevier.
11. Uscibooks.com. (2014). Stereochemistry. [online] Available at: https://www.uscibooks.com/Anslyn/AD06rt.pdf (accessed on 18 August 2020).

Chapter 3

Principles and Applications: Stereochemistry of Organic Compounds

CONTENTS

The chapter on the principle and the application of the stereochemistry of organic compounds, explains the significance of the molecular geometry as well as the chemical bonding that takes place in a molecule. This chapter also explains the significance of ionic bonding, the formation of ionic bonding, covalent bonding, and what is the difference between pure covalent bonds and polar covalent bond.

This chapter also provide highlights on the Lewis structure and Lewis symbols, along with the fundamentals of the double as well as triple bonds which includes discussions on formal charges and resonance

This chapter explains the strength of ionic bonds as well as the strength of the covalent bonds, which is consisting of bond strength, bond angle, bond length, and VSEPR theory. This chapter also addresses the significance of catenates, mechanically self-locked chiral gemini-catenanes, MIMs, and MSMs, as well as the molecular symmetry, chirality, and the effects of chirality on the molecular properties of a particular atom or molecule.

3.1. WHAT IS MOLECULAR GEOMETRY AND CHEMICAL BONDING?

Chemical bond is defined as the attraction between the atoms which permits the formation of the molecules and compound consisting of two or more atoms. The formation of the bond is caused by the electrostatic force of attraction in between atoms of different charges. The electrostatic force of attraction could take place either between electrons and nuclei (very strong), or as an outcome of a dipole attraction (weak).

All sorts of bonds can be described with the help of quantum theory, but, in the practice, simplification rules permit the anticipation of the strength, and directionality, along with the polarity of the bonds. The prime of examples are octet rule and VSEPR theory.

In addition to this, there is another and more detailed theory known as valence bond theory. Valence bond theory is consisting of orbital hybridization and resonance, along with the linear combination of the atomic or molecular orbital. Electrostatics can be applied in order to explain the polarities of the bonds and the impacts they have on the formation of chemical substances.

3.2. PRELUDE TO CHEMICAL BONDING AND MOLECULAR GEOMETRY

It has been acknowledged that the pure carbon take place in several numbers of various forms (which is known as allotropes). These forms of the pure carbon are consisting of graphite and diamonds as well. But it was not until the year of 1985 that a novel form of the carbon was determined: buckminsterfullerene, which is generally known as a "Buckyball."

There are several numbers of experimental evidence that exposed the formula, C_{60}, and then science experts found out that how sixty atoms of the carbon could form one symmetric, and a stable molecule by the help of the bonding theory. The bonding theory explains how the individual atoms connect to each other in order for more complicated structure. This is evident with respect to organic molecules and how these molecules form a stable aggregate of themselves.

3.2.1. Ionic Bonding

Atom either lose electrons or gain electrons, to achieve their ion form and have a stable electron configuration. The charges that are carried by the ions, which is either negative (anions) or positive (cations) determine their ability to either loose or gain electrons. Anions, or those that carry negative charges indicates how many electrons they can accept, while cations or those that carry positive charges indicates how many electrons they can lose or donate. These electrons therefore, are either lost or gained to enable the atom to have a stable outer shell configuration.

When a cation and anion combine, it forms a new substance called an ionic compound (salt). A salt, therefore is the product of a very strong electrostatic force between a cation and an anion. The very strong electrostatic bond in salts are called ionic bond. Atoms must totally transfer corresponding electrons to establish an ionic bond.

The properties or the uniqueness of the ionic compounds provides some lights on the nature of the ionic bonds. Ionic solids assume a crystalline structure which makes them rigid by nature. Some ionic solids are brittle while some have just right durability. Ionic bonds also tend to have both high melting point as well as high boiling point, making salts more durable by nature.

Ionic solids however, are not good electrical conductors, since the ions have no space to move freely in its solid state. Several ionic solids, never the less, dissolve readily in water.

Once the ionic solids dissolved or melted (in a solvent), the solution becomes excellent conductor of electricity as well as good conductor of heat. Solutions containing dissociated ions that are capable of conducting electricity are called electrolytes. The ability to conduct electricity is due to the random motion of negative and positive ions inside the solution.

Neutral atoms and the ions that are linked with them have very distinct chemical properties as well as physical properties. The atoms of the sodium, for instance, from the sodium metal, is a soft, silvery white metal which burns strongly when come in contact with air, and reacts violently (explosively) with the water. The atoms of the chlorine from the chlorine gas, Cl_2, a yellow-green gas is very corrosive with respect to most metals and it is very poisonous with respect to the animals as well as for plants. The strong reaction between the elements of the sodium and the chlorine forms the crystalline compound sodium chloride, which is white in color, and also known as the common table salt. The crystalline compound, sodium chloride is consisting of sodium cations and chloride anions.

The compound which is made up of these ions shows properties that are completely distinct from the properties of the both element sodium as well as chlorine. Chlorine is poisonous in nature, but sodium chloride is also very important for life. Similarly, the atoms of the sodium react strongly when they come in the contact of water, on the other hand, sodium chloride simply dissolves in water (Figure 3.1).

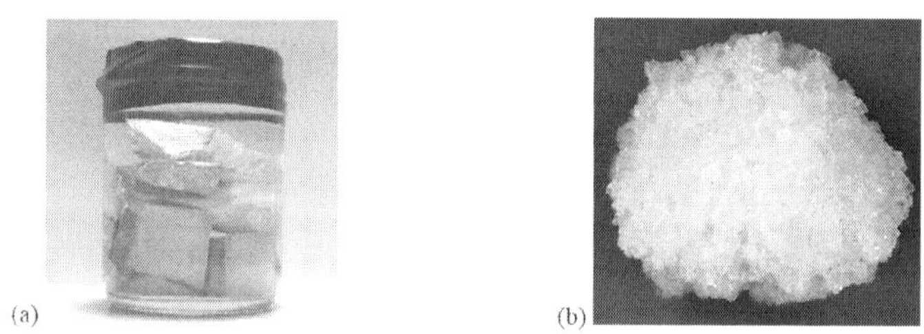

Figure 3.1. (a) Sodium is a soft metal that must be stored in mineral oil to prevent reaction with air or water. (b) When combined, they form white crystals of sodium chloride (table salt).

Source: Image (s) by Flickr.

3.2.2. The Formation of Ionic Compounds

The binary ionic compounds are made up of two elements. These two elements are, a metal (which helps in the formation of the cations) and the second element is a non-metal (which helps in the formation of the anions). For instance, NaCl or sodium chloride is a binary ionic compound. The ability of the substance to form binary ionic compound can also be predicted based on their periodic properties.

Several numbers of elements that have comparatively low ionization potential and lose electrons very easily. These are the elements that lie to the left of the periodic table.

Non-metal atoms have comparatively high electron attractions and readily gain electrons which are lost by metals. In this way, they fill their valence shells. The non-metallic elements are found in the upper right corner, with respect to the periodic table.

All of the substances (compounds) must be electrically neutral in nature. The total number of the positive charges on the cations of a compound must cancelled out by the total number of the negative charges that are present the anions, making the charge of the compound zero.

The formula of an ionic compound shows the simplest ratio of the numbers of ions that are important to give to the same numbers of positive as well as negative charges.

For instance, the formula for aluminum oxide, which is written as Al_2O_3, shows that this ionic compound has two Aluminum^{3+} which will require three oxide ions, O^{2-} to balance the charge of the compound; by the charge balance formula{(2 × +3) + (3 ×–2) = 0], we establish the charge of Al_2O_3 to be zero.

Nevertheless, it is very necessary to note that the formula for an ionic compound does not represent the physical arrangement of its ions. It is incorrect to use the term "molecule" for sodium chloride (NaCl) because there is not a single ionic bond, per se, between any specific pair of sodium and chloride ions.

The attractive forces between ions are isotropic in nature. Isotropic means that all the forces that are working is same in all directions. In simple words, any ion is equally attracted to all ions that are present nearby carrying the opposite charge.

This consequences in the ions spacing out themselves into a tightly bound, three-dimensional lattice structure. For instance, sodium chloride is consisting of a regular arrangement, which is having same amount of Na+ cations as well as Cl– anions (Figure 3.2).

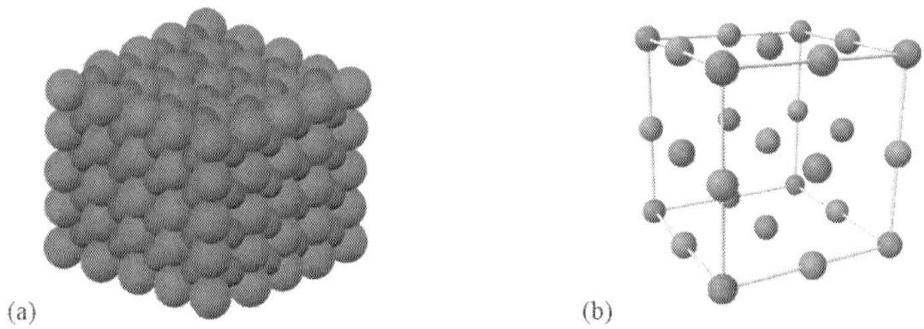
(a) (b)

Figure 3.2. The atoms in sodium chloride (common table salt) are arranged to (a) maximize opposite charges interacting. The smaller spheres represent sodium ions, the larger ones represent chloride ions. In the expanded view (b), the geometry can be seen more clearly. Note that each ion is "bonded" to all of the surrounding ions—six in this case.

Source: Image (a) by Needpix.com, Image (b) by Wikimedia commons.

3.2.3. Covalent Bonding

In ionic compounds, the electrons that are being transferred among the atoms of different elements in order to form an ion. But this is not the only way to do so or to form a compound. Atoms can also make chemical bonds with the help of sharing electrons between each other.

Ionic bonds occur because atoms have opposite charges. On the other hand, atoms of the same charges can also form bonds, called covalent bond. In this type of bond, however, electrons are not being transferred, but instead it is shared with the other atom. In some cases, atoms with the same electro negativity can share electrons and form diatomic of polyatomic substances.

For instance, two hydrogen atoms bond covalently in order to form H_2 molecule. In which each and every atom of the hydrogen that is present in the H_2 molecule has two electrons which helps in stabilizing the both hydrogen atom, giving each of the atom the equal count of valence electrons as the noble gas Helium (He).

It is well established that ionic bonds are stronger than covalent bonds due to the strong electrostatic attraction between ions. On the other hand, covalent bonds are held together by several secondary forces of attraction such as Van der Waals forces, dipole-dipole interaction, and hydrogen bonds. Generally, the compounds that bonded covalently have much lower melting and boiling points as compared to the compounds that are ionically bonded.

Several numbers of compounds in covalent bonds are liquids or gases at room temperature. On the other hand, majority of ionic compounds are in their solid states.

Ionic compounds are found to covalent the compounds are electrically neutral, and are poor conductors of electricity at any state.

3.2.4. Non-Polar vs. Polar Covalent Bonds

If the atoms that form a covalent bond are same in nature, just same as they are in hydrogen molecule H_2, and chloride molecule Cl_2, and other diatomic molecules, then the electrons in the bond must be shared in same amount. In this case, they are referred as a pure or non-polar covalent bond.

Electrons that are shared in a non-polar covalent bond have the equal probability of being near to each nucleus. In the case of molecule of chlorine Cl_2, each, and every atom that starts off with seven valence electrons, and each of the chlorine Cl atom shares one electron with the other, forming one covalent bond:

$$Cl + Cl \longrightarrow Cl_2$$

The total number of electrons around each individual atom is consisting of the six electrons that are nonbonding and two shared (that is bonding) electrons for eight total electrons, corresponding to the number of valence electrons in the noble gas argon, and satisfying conditions of the octet rule. Non-polar covalent bonds therefore, forms when two atoms of identical electronegativity shared their valence electrons.

When two atoms of different electronegativity forms, one of the atoms behave as a partially positive atom (less electronegative) and the one behaves as the partially negative atom (more electronegative), forming covalent bonds with polarity (or polar covalent bonds). In this case, there will be unequal electron density around the nucleus of the atoms, forming an unsymmetrical electron cloud.

For instance, in the formation of hydrochloric acid, chlorine is more electronegative than hydrogen. The hydrogen, thus will be the partially positive atom while chlorine is the partially negative atom. The electron cloud around chlorine therefore is bigger since it has the stronger tendency to attract electrons. In this case, the annotation in Figure 3 3, the polarity of the covalent bond is being represented by an arrow with a + charge on its origin aligned on top of the partially positive atom while the arrow head is aligned on top of the partially negative atom, indicating the direction of the net dipole.

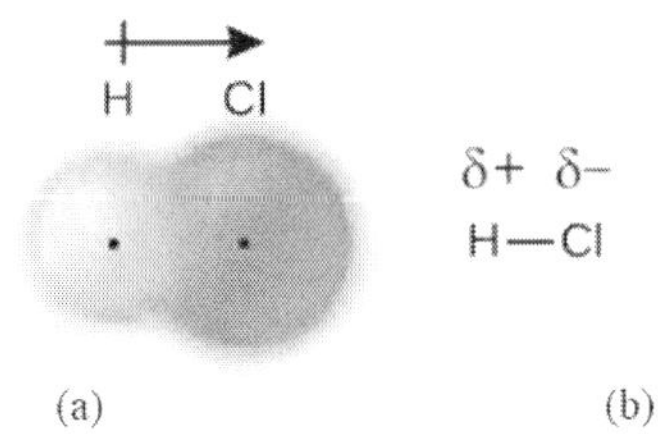

Figure 3.3. (a) The distribution of electron density in the HCl molecule is uneven. The electron density is greater around the chlorine nucleus. The small, black dots indicate the location of the hydrogen and chlorine nuclei in the molecule. (b) Symbols δ+ and δ–indicate the polarity of the H–Cl bond.

Source: Image by opentextbc.

3.3. LEWIS SYMBOLS AND STRUCTURES

Valence electron of an atom can be presented with the help of the drawing its Lewis symbols (with respect to the atoms as well as monatomic ions) and Lewis structures (with respect to the molecules and polyatomic ions).

Lone pairs, unpaired electrons, and single, double, or triple bonds are shown in the Lewis structure. Most of the structures, more particularly, those who are consisting of the second-row elements, obey the octet rule, in which every atom (apart from H) is surrounded with eight electrons.

Table 3.1 illustrates the Lewis symbols of some substances in their atomic form. On the other hand, in terms of molecules, bonding electrons are represented with lines. Only the lone pairs and unpaired electrons are retained around the atom. In an organic molecule, example a saturated hydrocarbon, the Lewis structure will exhibit no dots since all electrons are bonded (Table 3.2).

Table 3.1. Lewis Symbols Illustrating the Number of Valence Electrons for Each Element in the Third Period of the Periodic Table

Atoms	**Electronic Configuration**	**Lewis Symbol**
Sodium	[Ne] $3s^1$	Na •
Magnesium	[Ne] $3s^2$	• Mg •
Aluminum	[Ne] $3s^2$ $3p^1$	• Al • (one dot above)
Silicon	[Ne] $3s^2$ $3p^2$	• Si • (one dot above, one below)
Phosphorus	[Ne] $3s^2$ $3p^3$	• P • (two dots above, one below)
Sulfur	[Ne] $3s^2$ $3p^4$	: S • (two dots above, one below)
Chlorine	[Ne] $3s^2$ $3p^5$	: Cl • (two dots above, two below)
Argon	[Ne] $3s^2$ $3p^6$	: Ar : (two dots above, two below)

Source: Table from chem.libtexts.

Table 3.2. Cations are Formed When Atoms Lose Electrons, Represented by fewer Lewis Dots, Whereas Anions are Formed by Atoms Gaining Electrons. The Total Number of electrons Does Not Change

Metal	**Non–Metal**	**Ionic Compound**
Sodium atom	Chlorine atom	Sodium Chloride (Sodium ion and chlorine ion)
Magnesium atom	Oxygen atom	Magnesium Oxide (Magnesium ion and oxygen ion)

Calcium atom	Fluorine atom	Calcium fluoride (Calcium ion and two fluoride ions)

Source: Table from chem.libtexts.

3.3.1. Lewis Structures

Lewis symbols are used to represent the formation of covalent bonds, which are shown in Lewis structures. Lewis structures are the drawings which are used to explain the bonding in molecules as well as the polyatomic ions. For instance, when two atoms of the chlorine form a chlorine molecule, then one atom of chlorine share one pair of electrons to the other chlorine atom, as shown in the figure mentioned below:

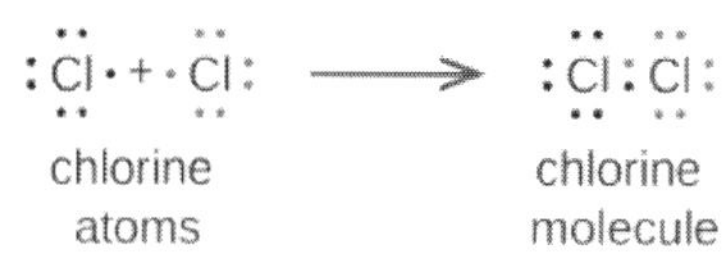

The Lewis structure represent that each and every atom of the chlorine Cl has three pairs of electrons that are not used in the formation of the bond (these unused pairs of the electrons are generally known as lone pairs) and one shared pair of electrons (written between the atoms). Generally, a dash (or a simple straight line) is used to represent the shared pair of electrons, as shown in the figure mentioned below:

With respect to the Lewis model, a single shared pair of electrons is a single bond. Each and every atom of the chloride Cl interrelates with the eight electrons total that are present in the valence shell. Out of which, the six in the lone pairs and the two in the single bond.

3.3.2. Double and Triple Bonds

As it has been mentioned above, when a pair of atoms shares one pair of electrons, it is known as a single bond. Nevertheless, a pair of atoms can require to share more than one pair of electrons in order to accomplish the mandatory octet.

The formation of the double take place, when two pairs of electrons are shared between a pair of atoms, just similar as they are shared between the carbon and oxygen atoms in CH_2O (formaldehyde) and just similar as that between the two carbon atoms in C_2H_4 (ethylene), as illustrated below:

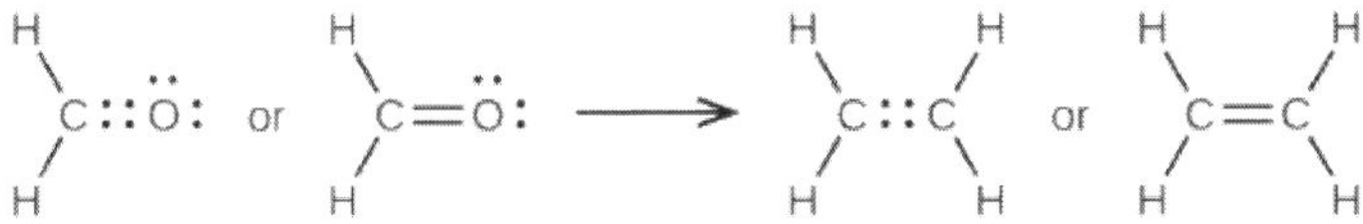

formaldehyde ethylene

A triple bond forms when three electron pairs are shared with the help of a pair of atoms, similar to that in carbon monoxide (CO) and the cyanide ion (CN–):

carbon monoxide cyanide ion

3.4. FORMAL CHARGES AND RESONANCE

With respect to the Lewis structure, formal charges can be allocated to each and every atom with the help of treating each bond as if one-half of the electrons are allocated to each and every atom. These sorts of hypothetical formal charges provide assistance in order to find out the most suitable Lewis structure.

A structure in which the formal charges are as close to zero as possible is preferred. Resonance take place in instances where two or more than two Lewis structures that are having same arrangements of atoms, but they have distinct allocation of the electrons can be written.

3.4.1. Calculating Formal Charge

The formal charge of the atom that is present in the molecule is the hypothetical charge that would have if it has been redistributed to the electrons in the bonds uniformly in between the atoms. Apart from this, the simpler way of saying this is that the formal charge results when the number of the valence electrons of the neutral atom, and subtracting the electrons that are non-bonding and then, subtracting the number of bonds that are linked with that atom which is present in the Lewis structure.

In this way, the calculation of the formal charge can take place, as shown in below:

Formal charge = # valence shell electrons (free atom) – # lone pair electrons – 1/2 # bonding electrons

The calculation of the formal charge can be verified with the help of identifying the sum of the formal charges for the whole structure. The sum of the formal charges of all atoms in a molecule must be zero; the sum of the formal charges in an ion should be equal to the charge of the ion.

It must be remembered that, the calculation of the formal charge for an atom is not the actual charge of the atom in the molecule. Formal charge is only a useful bookkeeping procedure; it does not represent the presence of actual charges.

3.5. STRENGTHS OF IONIC AND COVALENT BONDS

The strength of the bond explains that how strongly each and every atom of that particular molecule is linked to another atom, and in this way, how much energy is needed to break the bond between the two atoms.

This topic of the chapter will provide highlights regarding the bond strength of the covalent bonds, and then compared that to the strength of the ionic bonds, which is associated to the lattice energy of that particular compound.

3.5.1. Bond Strength: Covalent Bonds

The molecules that are stable exist because the covalent bonds clutch the atoms altogether. In order to gauge the strength of the covalent bond with the help of the energy that is needed in order to break it, that is, the energy which is important to separate the bonded atoms. Separating any sort of pair of bonded atoms, necessitates external energy. The stronger a bond, the more amount of energy is required in order to break the bond.

The amount of energy that is required to break the particular kind of covalent bond in one mole of the gaseous molecules is known as the bond energy, and it is also known as the bond dissociation energy. With respect to the diatomic molecule, the bond energy, D_{X-Y}, is defined as the change in standard enthalpy for the endothermic reaction, as illustrated below:

$$XY_{(g)} \longrightarrow X_{(g)} + Y_{(g)} D_{X-Y} = \Delta H^\circ$$

For instance, the bond energy of the pure covalent H–H bond, ΔH–H, is 436 kJ per mole of H–H bonds broken:

$H_{2(g)} \longrightarrow 2H_{(g)} D_{H\text{–}H} = \Delta H° = 436 kJ$

With respect to the molecules that are having three or more than three atoms have two or more than two bonds. The total of all bond energies in these kinds of molecules is equal to the standard enthalpy change, in the context of endothermic reaction that breaks all the bonds that are present in that particular molecule.

For instance, the total sum of the four C–H bond energies in CH_4, 1660 kJ, is equal to the standard enthalpy change of the reaction, the reaction is mentioned below:

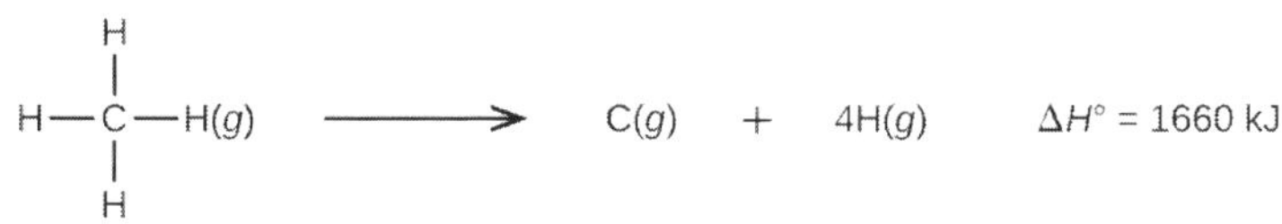

The average C–H bond energy, $D_{C\text{–}H}$, is 1660/4 = 415 kJ/mol because there are total four moles of C–H bonds broken per mole of the reaction. Never the less, the four C–H bonds are equivalent in the original molecule, individually they do not necessitate the same amount of energy in order to break the bond.

Once the first bond is broken (the amount of energy is required to break the bond is 439 kJ/mol), the remaining bonds are comparatively easier to break. The 415 kJ/mol value is the average, not the exact value of the energy which is needed in order to break any one bond.

As the number of the electron pairs in the bond increases, the strength of a bond between the two atoms increases. Typically, with the increment in the bond strength, the bond length reduces. In this way, it can be determined that the triple bonds are much stronger as well as shorter as compared to the double bonds between the same two atoms. In the same manner, the double bonds are stronger and shorter as well, when they are compared with the single bonds between the same two atoms.

Average bond energies for several numbers of bonds that are very common are shown in the table 3.3., mentioned below, along with the comparison of the bond lengths and the bond strength for some common bonds that are mentioned in Table 3.4. When one atom bonds to several numbers of atoms in a group, the bond strength are generally reduces as moving down in the group.

For instance, C–F is 439 kJ/mol, C–Cl is 330 kJ/mol, and C–Br is 275 kJ/mol.

Table 3.3. Bond Energies (kJ/mol)

Bond	Bond Energy	Bond	Bond Energy
C–S	260	**F–Cl**	255
C–Cl	330	**F–Br**	235
C–Br	275	**Si–Si**	230
C–I	240	**Si–P**	215
N–N	160	**Si–S**	225
N=NN=N	418	**Si–Cl**	359
N≡NN≡N	946	**Si–Br**	290
N–O	200	**Si–I**	215
N–F	270	**P–P**	215
N–P	210	**P–S**	230
N–Cl	200	**P–Cl**	330
N–Br	245	**P–Br**	270
O–O	140	**P–I**	215
O=OO=O	498	**S–S**	215
O–F	160	**S–Cl**	250
O–Si	370	**S–Br**	215
O–P	350	**Cl–Cl**	243
O–Cl	205	**Cl–Br**	220
O–I	200	**Cl–I**	210
F–F	160	**Br–Br**	190
F–Si	540	**Br–I**	180
F–P	489	**I–I**	150
F–S	285	**H–H**	436
C–C	345	**H–C**	415
C=C	611	**H–N**	390
C≡C	837	**H–O**	464
C–N	290	**H–F**	569
C=N	615	**H–Si**	395

C≡N	891	**H–P**	320
C–O	350	**H–S**	340
C=O	741	**H–Cl**	432
C≡O	1080	**H–Br**	370
C–F	439	**H–I**	295
C–Si	360	**C–P**	265

Source: Table from chem.libtexts.

Table 3.4. Average Bond Lengths and Bond Energies for Some Common Bonds

Bond	**Bond Length (Å)**	**Bond Energy (kJ/mol)**
C–C	1.54	345
C=CC=C	1.34	611
C≡CC≡C	1.20	837
C–N	1.43	290
C=NC=N	1.38	615
C≡NC≡N	1.16	891
C–O	1.43	350
C=OC=O	1.23	741
C≡OC≡O	1.13	1080

Source: Table from chem.libtexts.

3.6. MOLECULAR STRUCTURE AND POLARITY

As discussed in the above section of the chapter, two-dimensional Lewis structure are used to represent the molecules. Nevertheless, molecular structure is actually three-dimensional, and it is very significant to be able to explain the molecular bonds with respect of their distances, angles, along with the relative arrangements in the space (Figure 3.4).

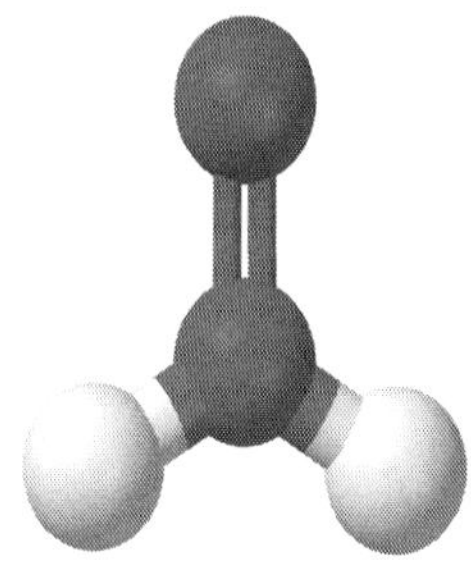

Figure 3.4. An illustration of formaldehyde molecule.

Source: Image by Wikimedia commons.

A bond angel is defined as, the angle between any kind of two bonds that are consisting of a common atom. Generally, the bond angle is measured in degrees. A bond length (or a bond distance) is the distance between the nuclei of two bonded atoms along the straight line which is connecting the nuclei. The bond distance is gauged in Angstroms (1 Å = 10–10 m) or picometers (1 pm = 10–12 m, 100 pm = 1 Å).

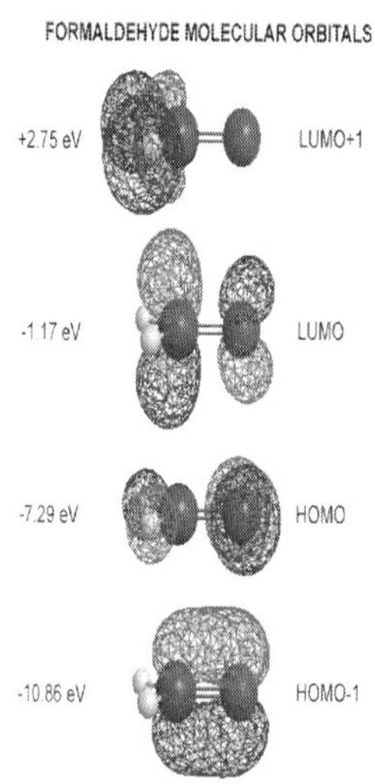

Figure 3.5. An illustration of formaldehyde molecule orbitals.

Source: Image by Flickr.

According to the VSEPR theory which anticipates the three-dimensional arrangement of the atoms inside a molecule. The VESPR theory states that the valence electrons will assume an electron-pair geometry that minimizes the repulsion between the regions of the high electron density (they could be bonds and/or lone pairs) (Figure 3.5).

Molecular structure refers only to the allocation of the atoms inside a molecule. Molecular structure does not refer to the placement of the electron, which is equal to the electron-pair geometry only when there are no lone electron pairs near around the central atom.

3.6.1. VSEPR Theory (Valence Shell Electron-Pair Repulsion Theory)

VSEPR theory, short form of Valence Shell Electron-Pair Repulsion theory, provide assistance in order to anticipate the molecular structure which is including the approximate bond angles around a central atom, of a molecule from an evaluation of the number of bonds and lone electron pairs in its Lewis Structure.

The valence shell electron-pair repulsion model assumes that the electron pairs that are present in the valence shell of the central atom will adopt an arrangement that minimizes the repulsion between these sorts of electron pairs with the help of maximizing the distance between them.

The electrons that are present in the valence shell of a central atom from either bonding pairs of the electrons, that are situated mainly between the bonded atoms, or lone pairs. The electrostatic repulsion of these sort of electrons is decreased when the several kinds of areas of high electron density assume the position or location as far from each other as possible.

The VSEPR theory, anticipates the arrangement of the electron pairs around each central atom and, generally, the suitable or appropriate arrangement of the atoms in the molecule. It should be acknowledged that, the theory only considers the electron-pair repulsions.

In addition to this, other interactions, for example, nuclear-nuclear repulsions and nuclear-electron attractions, are also take participation in the end or final arrangement that the atoms adopt in a particular molecular structure.

Taking the simplest case or simplest example of the VSEPR theory, let's assume the structure of a BeF_2 molecule which is present in gaseous state.

The Lewis structure of the BeF_2 molecule depicts only two electron pairs around the central beryllium atom.

With the two bonds and having no lone pairs of electrons on the central atom, the bonds are apart as far as possible. In addition to this, the electrostatic repulsion between these areas of high electron density is decreased to a minimum when they are on opposite sides of the central atom. The bond angle of the BeF_2 molecule is 180°, as shown below.

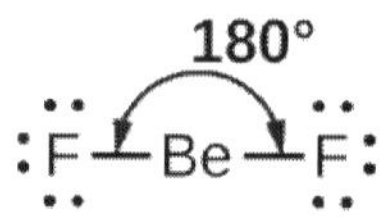

The BeF2 molecule adopts a linear structure in which the two bonds are as far apart as possible, on opposite sides of the Be atom.

3.7. CATENATES

Generally, catenates are macromolecules, which are mechanically interlocked to each other. Catenates can be fabricated with the help of the i-motif integration into several numbers of various deoxyribonucleic acid (DNA) assemblies, for example, nano circles that self-assemble into the nano rings.

With the help of the integration of the cytosine tracts into the DNA circles, the rings can be designed in order to undergo programmable contraction in the acidic conditions. These sorts of macrocyclic DNA rings work as a precursor with respect to the formation of the more complicated DNA topographies (Figure 3.6).

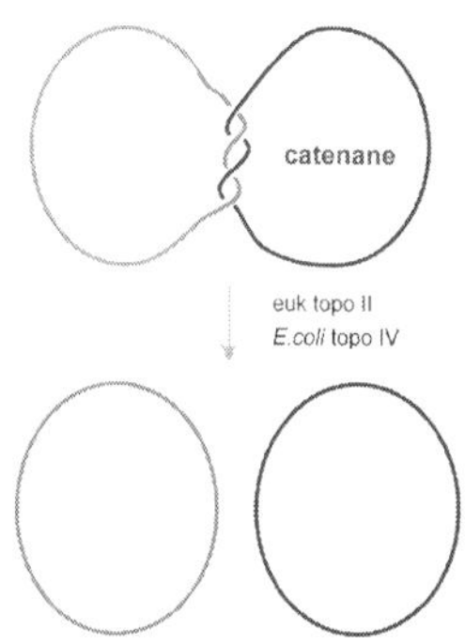

Figure 3.6. An illustration of catenates.

Source: Image by Wikimedia commons.

Together with rotaxanes and Borromean rings, catenates are interlocked DNA nanostructures that take place in nature and have been successfully synthesized. Catenates have been represented in DNA from HeLa cells and human leukemic leukocytes.

In the context of nature, catenates take participation in the action of to poisomerases in order to form knotting catenates, which will help in regulating the replication and genetic expression as well.

The process of synthesis of DNA catenates for the applications in DNA nanotechnology every so often, are consisting of the application of multiple ssDNA templates. These ssDNA templates are generally designed, in order to contain the regions of DNA complementarity.

When they are mixed in equal quantity the specificity of the hydrogen bonding encourages the DNA strands in order to self-assemble into an interlocking ring structure.

The 5′ end of each individual DNA strand is then covalently integrated to its 3′ end with the help of the enzymatic ligation. This 'locks' the DNA into the formation of the catenae, along with constantly stabilizing it.

Hopf links, which is singly-locked catenae (and, is generally consisting of two circles that are associated or connected together exactly once), and higher order linear, radial, and entwined catenates like Solomon links (which is consisting of two closed loops), which are twice as interlinked in an interlaced manner, and Borromean rings have all been described.

Because of their distinct ringed structure, the catenates have been used in several numbers of various DNA assemblies as well as sensors. In addition to this, they have been exploited in molecular machines as DNA switches and rotary motors.

Their use as switches is facilitated with the help of changing of the positions of the rings. On the other hand, the directionality or the rotation of the rings form the basis of their use as motors.

Because of the cavity that is present in the ring structure, that permits the housing of various sorts of molecules and the chemical moieties, the switching of the catenates can be accomplished with the help of a number of triggers, that are ranging from redox (CuI/CuII), cation, anion, and pH-induced designs.

The step of switching can also be facilitated with the help of sufficient amount of fuel/antifuel DNA strands. Notably, Lu, and co-workers have developed an olympiadane catenae nanostructure, in the face of the difficult tasks that are posed with the help of linking multiple rings.

From that point of time, same group has expanded their abilities in order to generate or develop the switchable, reconfiguring seven rings catenae nanostructure. A further design that uses the application of the catenates in the real-world applications is their incorporation onto the scaffolds or in order to help the networks.

3.7.1. Mechanically Self-Locked Chiral Gemini-Catenates

Entangled molecular structure as well as mechanically interlocked molecular architecture shows one of the elaborate topological superstructures that have been engineered at a molecular resolution. There are several methodologies or approaches present in order to fabricate the mechanically self-locked molecules (MSMs), with the help of effective one-step amination of a pseudo rotaxane which is made out from the dual functionalized pillar arene (P [5]A) threaded with the help of α, ω-diaminoalkane (DA-n; n=3–12).

In this way, both, the monomeric as well as dimeric pseudo catenates can be acquired. These pseudo catenates are inherently chiral and the reason behind this is the topology of P[5]A that is being used, were isolated and completely categorized with the help of NMR and circular dichroism spectroscopy, X-ray crystallography, and density-functional theory (DFT) calculations.

In addition to this, the dimeric pseudo catenae, which is known as 'Gemini-catenae,' is consisting of stereoisomeric meso-erythro and dl-threo isomers. Out of which two P[5]A moieties are threaded with the help of two DA-n chains in topologically different patterns.

This leads to the chiral pseudo catenates as well as Gemini-catenates that will greatly encourage the practical application of these sorts of sophisticated chiral architectures in supramolecular and materials science and technology.

Molecules that are interlocked mechanically along with the entangles molecular architectures have attracted much interest because of their interesting topological aspects and characteristics as well as the increment in the impacts of the material science along with the nanotechnology.

Partly stimulated with the help of mathematical, iconological as well as religious symbols, the progress has been made in the rational synthesis of the mechanically interlocked molecules (MIMs) with ever-growing complicatedness.

An extensive variety of both bi-as well as multi-component MIMs, which is consisting of [n] rotaxanes (n≥2), [n] catenates (n≥2) and [n]daisy chains (n≥2), have hitherto been synthesized.

On the other hand, the covalently associated single-component locked molecules, which are known as MSMs, have intermittently been reported. Common examples of MSMs include pseudo[1] rotaxane, pseudo [1] catenae, pretzelane, and double-lasso rota macro-cycle.

When they are compared with the well-established MIMs, never the less, the development of higher-order mechanically self-locked molecule systems still remains a difficult task, and the main reason behind this that the deficiency of a convenient general synthetic route and the tremendous difficulties in both synthesis as well as characterization.

1-Hydroxybenzotriazole (HOBT), which is an extensively used additive in the peptide synthesis, has also been make into use in order perform the synthesis of cyclic peptides, which stimulated to use this reagent in the synthesis of pseudo[1] catenates as well as single-component MSMs with the help of amination.

In this synthesis, HOBT is applied in order to perform the synthesis of a sequence of MSMs. And, it has been found that the amination mediated with the help of HOBT (also known as 1-hydroxybenzotriazole) serves as the important step in order to develop or formulate these sophisticated topologies.

Fascinatingly, when a long α,ω-diaminoalkane was reacted with pillar [5] arene-dicarboxylic acid in the presence of a condensation reagent 1-ethyl-3-(3-dimethylaminopropyl)carbodiimide (EDCI) and HOBT, a new member of the MSM family was obtained and eventually identified as a dually threaded pseudo [1] catenae, or 'Gemini-catenae' in addition to a singly threaded pseudo [1] catenae, both of which are conformationally stable and inherently chiral.

3.8. MOLECULAR SYMMETRY

In several numbers of different cases, the symmetry of a molecule offers a great deal of information with respect to its quantum states, even without a thorough or a comprehensive solution of the Schrödinger equation.

A geometrical transformation which changes or transforms a molecule into an indistinguishable copy of itself, this process is known as symmetry operation. A symmetry operation can be consisting of a rotation about an axis, an inversion through a point, a reflection in a plane, or some combination of these.

3.8.1. The Ammonia Molecule

In any symmetry operation on NH_3, the nitrogen atom remains fixed, but the hydrogen atoms can be permuted in 3!=6 various ways. The axis of the molecule is known as a C_3 axis. Since the molecule can be rotated about it into three equivalent orientations, 120° apart. Generally speaking, a C_n axis has 'n' equivalent orientations, separated with the help of 2π/n radians.

The axis of highest symmetry in a molecule is known as the principal axis. There are three mirror planes, which are designated as $\sigma_1, \sigma_2, \sigma_3$, run with the help of the principal axis in ammonia. These are designated as σ_v or vertical planes of symmetry. Ammonia fits to the symmetry group designated C_{3v}, which is categorized with the help of a three-fold axis with three vertical planes of symmetry.

Let's allocate the orientation of the three hydrogen atoms as shown in figure below as {1, 2, 3}, reading in clockwise order from the bottom. A counterclockwise rotation by 120°, allocated with the help of the operator named C_3, which developed the orientation {2, 3, 1}.

A second counterclockwise rotation, allocated $C_2^{\,3}$, which develops the orientation of {3, 1, 2}. Considering that two successive counterclockwise rotations by 120° is equivalent to one clockwise rotation by 120°. In the same way, the last operation could also be designated C_{-13}.

The three reflection operations σ_1, σ_2, σ_3, applied to the original configuration {1, 2, 3}, which as an outcome produces {1, 3, 2}, {3, 2, 1} and {2, 1, 3}, respectively. In the end, it must include the identity operation, which is designated by E, which leaves an orientation unaffected. The impacts of the six possible operations of the symmetry group C_{3v} can be summarized as follows:

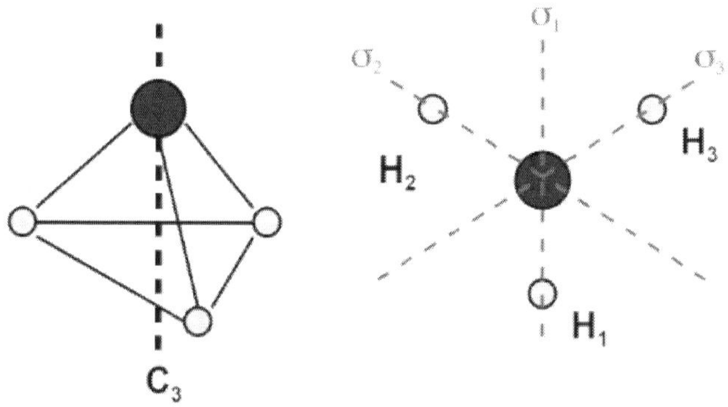

To view of ammonia molecule,

$E \{1,2,3\} = \{1,2,3\}$ $C_3 \{1,2,3\} = \{2,3,1\}$

$C_{23} \{1,2,3\} = \{3,1,2\}$ $\sigma_1\{1,2,3\} = \{1,3,2\}$

$\sigma_2 \{1,2,3\} = \{3,2,1\}$ $\sigma_3 \{1,2,3\} = \{2,1,3\}$

In this way, it is accounted for all 6 possible permutations of the three hydrogen atoms.

The successive application of two symmetry operations is equivalent to some single symmetry operation. For example, applying C_3, then σ_1 to our starting orientation, we have

$\sigma_1 C_3 \{1,2,3\} = \sigma_1 \{2,3,1\} = \{2,1,3\}$

But this is equivalent to the single operation σ_3. This can be represented as an algebraic relation among symmetry operators.

Note that successive operations are applied in the order right to left when represented algebraically. For the same two operations in reversed order, we find

$C_3\sigma_1\{1,2,3\} = C_3\{1,3,2\} = \{3,2,1\} = \sigma_2\{1,2,3\}$

In this way, symmetry operations do not, in general commute

$AB \neq BA$

Never the less, they may commute, for example, C_3 and C_2^3.

The algebra of the group C_{3v} can be summarized by the following multiplication table.

1st	E	C_3	C_2^3	σ_1	σ_2	σ_3
2nd						
E	E	C_3	C_2^3	σ_1	σ_2	σ_3
$\mathbf{C_3}$	C_3	C_2^3	E	σ_3	σ_1	σ_2
$\mathbf{C_2^3}$	C_2^3	E	C_3	σ_2	σ_3	σ_1
$\boldsymbol{\sigma_1}$	σ_1	σ_2	σ_3	E	C_3	C_2^3
$\boldsymbol{\sigma_2}$	σ_2	σ_3	σ_1	C_2^3	E	C_3
$\boldsymbol{\sigma_3}$	σ_3	σ_1	σ_2	C_3	C_2^3	E

Notice that each operation occurs once and only once in each row and each column.

3.8.2. Group Theory

With respect to the mathematics, a group is defined as a set of g elements $G \equiv \{G_1, G_2. G_h\}$ together with a rule for combination of elements, generally, which is refer to as a product. The elements must fulfil the following four conditions.

The product of any two elements of the group is another element of the group. That is $G_iG_j=G_k$ with $G_k \in G$.

Group multiplication must fulfil an associative law, $G_i(G_jG_k)=(G_iG_j) G_k \equiv G_iG_jG_k$

There exists an identity element E such that $EG_i=G_iE=G_i$ for all i.

Every element G_i has a distinct inverse $G-1_i$, such that $G_iG-1_i=G-1_iG_i=E$ with $G-1_i \in G$.

The number of elements h is called the order of the group. In this way, C_{3v} is a group of order six.

A group of quantities which fulfill the group multiplication table is called a representation of the group. Due to the possible noncommutativity of group elements, simple numbers are not always satisfactory with respect to the represent groups, hence, every so often application of matrices comes into the play.

The group C_{3v} has three irreducible representations, or IR's, which cannot be broken down into simpler representations. A trivial, but, never the less important, representation of any sort group is the entirely symmetric representation, in which each group element is represented by 1. The multiplication table then simply reiterates that $1 \times 1=1$.

For C_{3v} this is called the A_1 representation:

A_1: $E=1, C_3=1, C_2^3=1, \sigma_1=1, \sigma_2=1, \sigma_3=1$

A slightly less trivial representation is A2:

A2:$E=1, C_3=1, C_2^3=1, \sigma_1=-1, \sigma_2=-1, \sigma_3=-1$

The operations C_3 and C_{23} are said to belong to the same class, from the time they perform the identical geometric function, but for various orientations in space. Analogously, σ_1, σ_2 and σ_3 are obviously in the same class. E is in a class by itself.

The class structure of the group is designated with the help of $\{E, 2C_3, 3\sigma_v\}$. Another important theorem states that the total of the squares of the dimensionalities of the irreducible representations of a group adds up to the order of the group.

3.9. CHIRALITY AND SYMMETRY

All objects can be categorized with respect to a property, which is known as chirality (from the Greek "cheir" meaning hand). A chiral object is not symmetrical in all respects (that is, it is superimposable) with its mirror image. An achiral object is identical with (super imposable on) its mirror image.

Chiral objects have a "handedness," for instance, a corkscrew, scissors, golf clubs, and shoes. In this way, one can buy right or left-handed golf clubs and scissors. In the same manner, gloves, and shoes come in pairs, a right and a left.

Achiral objects do not have a handedness, for instance, a baseball bat (no writing or logos on it), a plain round ball, a pencil, a T-shirt and a nail. The chirality of any kind of material or an object is linked to its symmetry, and to this end it is useful in order to find out the certain symmetry elements that can be linked with a given object.

A symmetry element is a plane, a line or a point in or with the help of an object, about which a rotation or reflection leaves the object in an orientation indistinguishable from the original. Some examples of symmetry elements are shown in Figure 3.7.

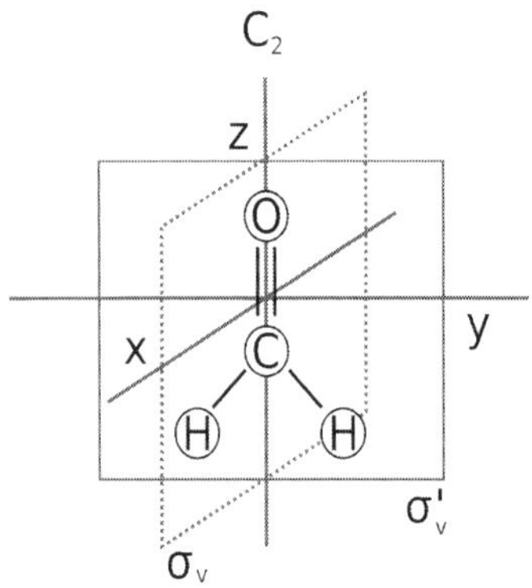

Figure 3.7. An illustration of molecular symmetry and chirality.

Source: Image by Wikipedia.

The symmetry of a molecule explains how its various parts relate to one another geometrically. Symmetry plays a very significant role in several numbers of fields of chemistry, with effects on:

1. Physical properties. For example, dipole moment, chirality;
2. Spectroscopic properties: transition intensities, geometric equivalence of groups or nuclei; and
3. Bonding interactions: bonds necessitate overlap of atomic orbitals of correct symmetry

3.9.1. Terms and Definitions

1. **Symmetry Operation:** a spatial manipulation which is performed on a molecule that leaves it in a configuration identical to and super imposable upon the original configuration.
2. **Symmetry Element:** an axis, plane, or point about which a symmetry operation is performed.
3. **Point Group:** a symbol that shows all the symmetry elements present in a molecule.

3.9.2. Symmetry Elements and Operations

Element	Operation	Symbol	Notes
none	Identity (no operation)	E	1. a molecule is always unchanged if no operation is performed, so all molecules possess E 2. $E = C_1$ i.e., rotation by 360° about any axis is just like doing nothing
proper rotation axis	rotate by 360°/n	C_n	1. axis with highest value of n is the principal axis 2. C_2 axes perpendicular to the principal C_n are called C_2' axes
mirror plane	Reflection	σ	1. σ parallel to and containing C_n is vertical: σv 2. σ parallel to C_n and bisecting two C_2' axes is dihedral: σ_d 3. σ perpendicular to Cn is horizontal: σ_h
inversion center	inversion	i	1. all atoms are moved through inversion center to an equal distance on opposite side 2. it may or may not be an atomic center
improper rotation axis	rotate by 360°/n, then reflect ^ to axis	S_n	1. combination rotation/reflection operation 2. odd n value requires both C_n and σ_h (e.g., BF_3 has an S_3 axis, and also has both a C_3 and a σ_h) 3. even 'n' value may or may not have C_n and σ_h (e.g., CH_4 has an S_4 axis, but has neither a C_4 nor a σ perpendicular to the S_4) 4. $S_1 = \sigma$, $S_2 = i$

3.9.3. Effects on Molecular Properties

1. Chirality:

In order to be chiral, a molecule must lack both i and σ.*

Molecules belonging to groups C_n (including E) and D_n are chiral.

* This is a small oversimplification. The most detailed or detailed condition for chirality is the deficiency of any sort of S_n element. In addition to this, because hardly anybody really knows what those look like, there are a sequence of more and more precise shortcuts that are used to Spotting a Chiral Compound.

2. Polarity:

In order to be polar, a molecule must lack all of i, C_2' axes, and σ_h.

Molecules belonging to groups C_s, C_n (including E) and C_{nv} (including $C_{\infty v}$) can be polar.

3.10. CONCLUSION

The study of the chiral molecules is an important branch of stereochemistry. Stereochemistry of the organic compounds is consisting of the methodologies or tactics that are being used in order to identify and explain the relationships.

It is very important to understand the how the chirality impacts the chemical reactions. The molecular symmetry also impacts the chirality and the spatial arrangements of the molecules involved in the chemical reaction. This is very important to understand as the symmetry operations taking place explain how these atoms behave differently when they are rotated at different angles. Chirality and symmetry play a very important role in the major applications of stereochemistry.

REFERENCES

1. *(n.d.). Chemical Bonding and Molecular Geometry.* [ebook] pp. 345–391. Available at: https://web.ung.edu/media/chemistry/Chapter7/Chapter7-ChemicalBonding-MolecularGeometry.pdf (accessed on 18 August 2020).
2. Chemeddl, (n.d.). *Definitions: Chiral.* [online] Chemeddl.org. Available at: http://www.chemeddl.org/resources/stereochem/definitions10.htm (accessed on 18 August 2020).
3. Chemistry LibreTexts, (2019). *11: Molecular Symmetry.* [online] Chemistry LibreTexts. Available at: https://chem.libretexts.org/Bookshelves/Physical_and_Theoretical_Chemistry_Textbook_Maps/Supplemental_Modules_(Physical_and_Theoretical_Chemistry)/Quantum_Mechanics/11%3A_Molecules/11%3A_Molecular_Symmetry (accessed on 18 August 2020).
4. Chemistry LibreTexts, (2019). *7.0: Prelude to Chemical Bonding and Molecular Geometry.* [online] Chemistry LibreTexts. Available at: https://chem.libretexts.org/Courses/University_of_Kentucky/UK%3A_General_Chemistry/07%3A_Chemical_Bonding_and_Molecular_Geometry/7.0%3A_Prelude_to_Chemical_Bonding_and_Molecular_Geometry (accessed on 18 August 2020).
5. Chemistry LibreTexts, (2019). *7.1: Ionic Bonding.* [online] Chemistry LibreTexts. Available at: https://chem.libretexts.org/Courses/University_of_Kentucky/UK%3A_General_Chemistry/07%3A_Chemical_Bonding_and_Molecular_Geometry/7.1%3A_Ionic_Bonding (accessed on 18 August 2020).
6. Chemistry LibreTexts, (2019). *7.2: Covalent Bonding.* [online] Chemistry LibreTexts. Available at: https://chem.libretexts.org/Courses/University_of_Kentucky/UK%3A_General_Chemistry/07%3A_Chemical_Bonding_and_Molecular_Geometry/7.2%3A_Covalent_Bonding (accessed on 18 August 2020).
7. Chemistry LibreTexts, (2019). *7.3: Lewis Symbols and Structures.* [online] Chemistry LibreTexts. Available at: https://chem.libretexts.org/Courses/University_of_Kentucky/UK%3A_General_Chemistry/07%3A_Chemical_Bonding_and_Molecular_Geometry/7.3%3A_Lewis_Symbols_and_Structures (accessed on 18 August 2020).

8. Chemistry LibreTexts, (2019). *7.4: Formal Charges and Resonance*. [online] Chemistry LibreTexts. Available at: https://chem.libretexts.org/Courses/University_of_Kentucky/UK%3A_General_Chemistry/07%3A_Chemical_Bonding_and_Molecular_Geometry/7.4%3A_Formal_Charges_and_Resonance (accessed on 18 August 2020).
9. Chemistry LibreTexts, (2019). *7.5: Strengths of Ionic and Covalent Bonds*. [online] Chemistry LibreTexts. Available at: https://chem.libretexts.org/Courses/University_of_Kentucky/UK%3A_General_Chemistry/07%3A_Chemical_Bonding_and_Molecular_Geometry/7.5%3A_Strengths_of_Ionic_and_Covalent_Bonds (accessed on 18 August 2020).
10. Chemistry LibreTexts, (2019). *7.6: Molecular Structure and Polarity*. [online] Chemistry LibreTexts. Available at: https://chem.libretexts.org/Courses/University_of_Kentucky/UK%3A_General_Chemistry/07%3A_Chemical_Bonding_and_Molecular_Geometry/7.6%3A_Molecular_Structure_and_Polarity (accessed on 18 August 2020).
11. Chemistry LibreTexts, (2019). *7: Chemical Bonding and Molecular Geometry*. [online] Chemistry LibreTexts. Available at: https://chem.libretexts.org/Courses/University_of_Kentucky/UK%3A_General_Chemistry/07%3A_Chemical_Bonding_and_Molecular_Geometry (accessed on 18 August 2020).
12. Chemistry LibreTexts, (2019). *Chirality and Symmetry*. [online] Chemistry LibreTexts. Available at: https://chem.libretexts.org/Bookshelves/Organic_Chemistry/Supplemental_Modules_(Organic_Chemistry)/Chirality/Stereoisomers/Chirality_and_Symmetry (accessed on 18 August 2020).
13. Connolly, A., Seow, N., Fenati, R., & Ellis, A., (2019). DNA Nanostructures. *Comprehensive Nanoscience and Nanotechnology*, [online] pp. 1–26. Available at: https://www.sciencedirect.com/science/article/pii/B978012803581810445X (accessed on 18 August 2020).
14. Li, S., Zhang, H., Xu, X., & Liu, Y., (2015). Mechanically self-locked chiral Gemini-catenates. *Nature Communications,* [online] *6*(1). Available at: https://www.nature.com/articles/ncomms8590 (accessed on 18 August 2020).

15. Science Direct, (2014). *Catenane - An Overview | Sciencedirect Topics.* [online] Sciencedirect.com. Available at: https://www.sciencedirect.com/topics/chemistry/catenane (accessed on 18 August 2020).
16. Stephen, M. W., (2010). *Molecular Symmetry and Chirality - How to Spot a Chiral Compound Part I.* [online] People.ok.ubc.ca. Available at: https://people.ok.ubc.ca/wsmcneil/symm/chirali.htm (accessed on 18 August 2020).

Chapter 4 Conformation of Acyclic and Cyclic Molecules

CONTENTS

This chapter discusses about the basic chemistry of biomolecules, namely carbohydrates, proteins, lipids, and nucleic acids. Along with the discussions of their primary structures, functions, and conformational configurations, their importance as well as uses onto the living cell and participation into the dynamic biochemical system will also be tackled.

This chapter also addresses the several kinds of homocyclic compounds such as aromatic compounds (benzenoid), heterocyclic compounds (or non-benzenoid aromatics), general aspects of heterocyclic compounds, and comparison with carbocyclic compounds.

The conformation of acyclic molecules which includes conformation of ethane, conformation of butane is also explained and provide highlights in the conformation of cycloalkanes or cyclic compounds

4.1. INTRODUCTION

Organic compounds are a large class of the chemical compounds in which one atom or more carbon atoms are connected by covalent bonds to other atoms. In most cases, this takes place in hydrogen, oxygen, and nitrogen. Some or very few compounds that are consisting of carbon are not categorized as the organic which includes carbides, cyanides, and carbonated.

Living things are mostly organic compounds. And the reason behind this is organism's association with carbon which is the basic requirement of all living cell.

The organic compounds are linked with the life process, are the one of the important scopes organic chemistry. These molecules performing basic functions for the living cells are characterized to be large organic molecules or often called "macromolecules. The four macromolecules deemed essential to the living cell and the biochemical processes of the living organism are:

1. Carbohydrates;
2. Lipids;
3. Proteins; and
4. Nucleic acids.

4.2. WHAT ARE CARBOHYDRATES?

Nearly all of the organisms use carbohydrates as the source of the energy. In addition to this, some of the carbohydrates functions as the structural materials. In simple words, carbohydrates are basically molecules that are made up of carbon, hydrogen, and oxygen. The ratio of carbon to hydrogen to oxygen atom in carbohydrates is 1:2:1, and carbohydrate molecular formula obey the general equation $C_nH_{2n}O_n$, where n corresponds to the number of carbon atoms. Carbohydrates are sometimes referred to as sugars.

The simplest form or monomers of carbohydrates are the monosaccharides, which is a single sugar molecule, they are then classified as disaccharides, if they are made up of two molecules of sugar, trisaccharides for 3 units of sugar and polysaccharides for more than 3 units. The most important or most significant monosaccharide is glucose. Glucose is a carbohydrate with the molecular formula of $C_6H_{12}O_6$.

Glucose is the basic form of the fuel that are present in the living organisms. Glucose is water soluble and can be transported through the body and to each cell with the help of body fluids, where it is metabolized in order to for the body to harness the energy. Glucose is the starting material for the cellular respiration, and it is the main product of the photosynthesis. Majority of complex and simple sugars are metabolized to glucose as the major form of absorbable sugar by the blood stream. It is also one of the most measured parameters in the blood and is a starting indicator for the diagnosis of diabetes.

There are three main disaccharides that are also found among living organisms, namely; maltose, sucrose, and lactose. Maltose is a combination of the two glucose units that are linked with the help of covalent bonds. Table sugar is also known as sucrose is composed of 1 unit of glucose and 1 unit of fructose, while lactose (the sugar in milk (is made up of glucose and galactose.

Figure 4.1 shows that water is generated during the synthesis of sucrose through a process known as dehydration reaction. The reversal of the process is hydrolysis. Hydrolysis is the process in which the molecule is split and water is added.

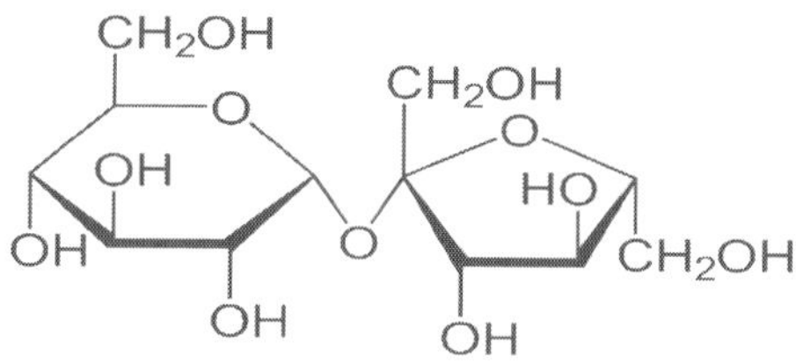

Figure 4.1. Glucose and fructose molecules combine to form the disaccharide sucrose.

Source: Image by Wikipedia.

Polysaccharides are made up with linking (polymerization) of monosaccharides (usually more than 3 units). Out of which, one of the most important or most significant polysaccharides is starch. Starch is composed hundreds or thousands of the units of glucose that are associated with each other through a specialized linkage for carbohydrates or what we call glycosidic bonds.

Starch is the major energy source of many living organism. Foods that are rich in carbohydrates such as rice, wheat, corn, and potatoes are the staple foods of many living organisms due to its ability to fulfill the body's energy requirement (Figure 4.2).

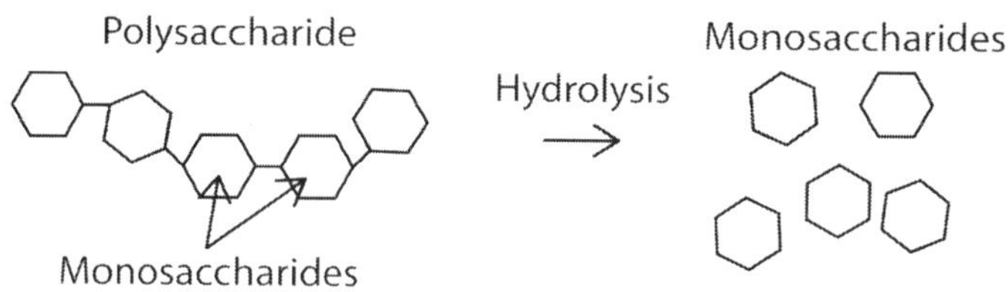

Figure 4.2. A simple polysaccharides hydrolysis.

Source: Image by Wikimedia Commons.

Another important polysaccharide, are glycogen and cellulose. Starch, similar to glycogen is had monosaccharides liked together by the α-1,4-glycosidic bonds which makes it a compatible energy storage molecule among animals and is stored in the liver. On the other hand, cellulose is a structural component of plant cell walls and its monosaccharides are linked together by the β-1,4-glycosidic bonds. Thus, it's commonly said that glycogen is starch associated with animals while cellulose is the starch associated with plants.

Wood is mainly made up of cellulose, along with the plant cell walls. Cotton fabric as well as paper are commercial products that are also made up of cellulose.

4.3. WHAT ARE LIPIDS?

Commonly, lipids are organic molecules that are composed of carbon, hydrogen, and oxygen. The ratio of the hydrogen atoms to the atoms of the oxygen is much higher in lipids, when they are compared to the ratio of hydrogen atoms to the oxygen atoms in carbohydrates. Lipids are also considered as the backbone of steroids (steroids are the material of which several numbers of hormones are made up of), waxes, and fats.

The molecules of fats are composed of glycerol molecule with one, two or three molecules of fatty acids. A glycerol molecule is consisting of three hydroxyl (–OH) groups. A fatty acid is a long chain hydrocarbon (near about four to twenty-four), having a carboxyl (–COOH) group at one end.

The fatty acids in fats may all be similar to each other, or they may all be different to each other as well. Fatty acids are bound to the glycerol molecules with the help of a process that is consisting of the removal of the water (Figure 4.3).

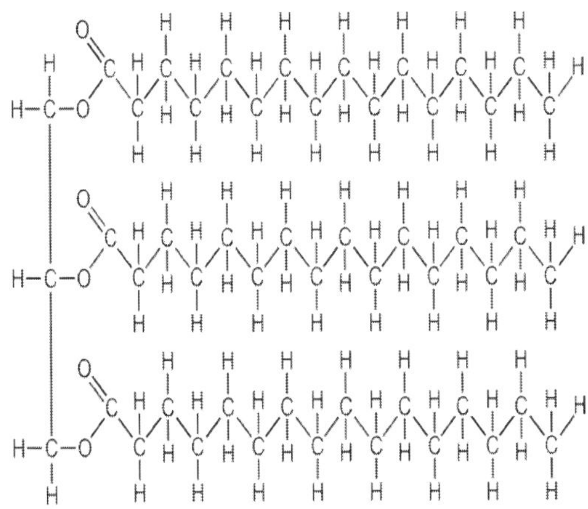

Figure 4.3. An illustration of fat molecule structure: Fat molecule is constructed by combining a glycerol molecule with three fatty acid molecules.

Source: Image by Free SVG.

There are specific types of fatty acids that are having one double bond or more than one double bond in their molecules. Fats which is consisting of these molecules are the unsaturated fats. Other fatty acids have no double bonds. Fates that are consisting of these sorts of fatty acids are saturated fats in nature. in several numbers of cases of human health situation, the utilization of the unsaturated fats is preferred to the utilization of the saturated fats.

Fats that are stored in the cells are generally form clear droplets of oil, which are known as globules. And the behind is that, fats do not dissolve in water. every so often, plants store fats inside their seeds, animals store the fats in large, clear globules in the cells of the adipose tissue.

The fat which is present in the adipose tissues, is consisting of much concentrated energy. In this way, they play a role as a reserve energy supply with respect to the organism. The enzymes lipase breaks down the fat into the fatty acids as well in glycerol inside the digestive system of the human being.

4.4. WHAT ARE PROTEINS?

Out of all complicated organic compounds, the most complicated organic compound, which is known as proteins. Proteins are made up with the help of amino acids. Amino acids or proteins are consisting of atoms of hydrogen, atoms of oxygen, atoms of carbon, along with the atoms of nitrogen as well. there are particular amino acids that are having some content of atoms of sulfur, atoms of phosphorus, along with the other trace elements such as copper and iron (Figure 4.4).

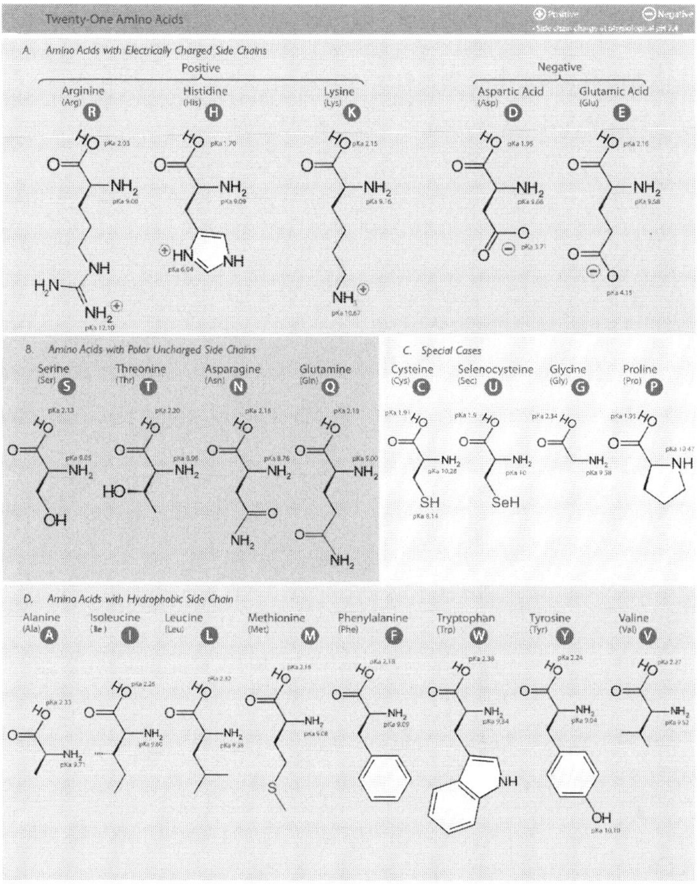

Figure 4.4. The structure and chemistry of amino acids: An illustration of molecular structure of 21 proteinogenic amino acids.

Source: Image by Wikimedia Commons.

There are several numbers of proteins that are immensely and extremely complicated in nature. Nevertheless, all sorts of proteins are made up with the help of long chains of comparatively simple amino acids. There are twenty sorts of amino acids exists. Each and every amino acid has an amino ($-NH_2$) group, a carboxyl (–COOH) group, and a group of atoms called an –R group (where R stands for radical).

The amino acid fluctuates according to the nature of the –R group. Some examples of the amino acids are alanine, valine, glutamic acid, tryptophan, tyrosine, and histidine.

The removal of the molecules of water are connected to the amino acids in order to form a protein. The process protein formation is embedded in the dehydration synthesis, and a by-product of this process or this synthesis is water. The links forged between the amino acids are peptide bonds, and small proteins are every so often known as peptides (Figure 4.5).

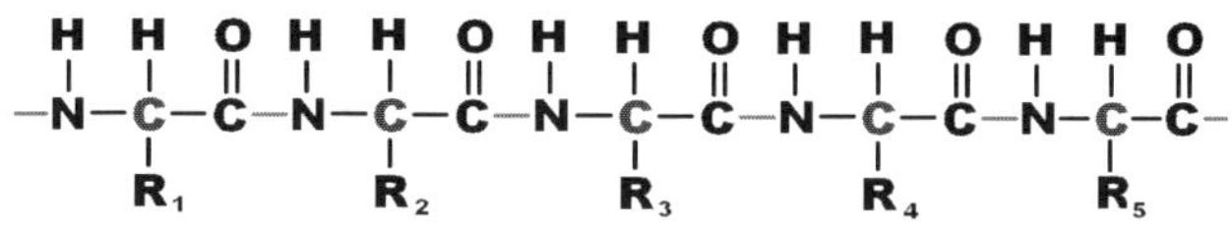

Figure 4.5. An illustration of simplified peptide chain.

Source: Image by Wikimedia Commons.

Each and every living being or living organism rely on the proteins for their existence. Proteins are the most important molecules from which living organisms are made up of. There are specific types of proteins that are dissolved or suspended in the watery substance of the cells. On the other hand, others are combined or integrated into several numbers of various structure of the cells.

Proteins are also found as supporting as well as strengthening materials that are present in the tissues outside the cells. Bone, cartilage, tendons, as well as ligaments are all made up with the help of proteins.

One more important function of the protein is an enzyme. The fundamental role of the enzymes is to speed up the chemical reaction that take place inside the cells. Enzymes are not used up in the reaction, as an alternative, the enzymes remain available in order to catalyze or speed up the succeeding reactions.

Each and every species forms the proteins distinct with respect to that particular species. The information for synthesizing the distinct protein is situated in the nucleus of the cell. The so-called genetic code shows the sequence or series of the amino acid in the proteins.

In this way, the genetic code shows the chemistry that take place inside a cell. Proteins also can serve as a reserve source of energy for the cell. On the other hand, the amino acids group is removed or eradicated from an amino acid, the outcome compound is full of energy.

4.5. WHAT ARE NUCLEIC ACIDS?

Similar to the proteins, the nucleic acids are also very large molecules in nature. The nucleic acids are made up with the help of smaller units, and these smaller units are known as nucleotides. Each and every nucleotide is consisting of the molecule of carbohydrate (generally known as sugar), a group of phosphate, and a nitrogen, which is containing molecule that, because of its properties, is a nitrogen base.

With respect to the all living organisms, they have two important or significant nucleic acids. One kind of nucleic acid is known as deoxyribonucleic acid, or DNA. On the other hand, the second kind of the nucleic acid is known as ribonucleic acid, or RNA.

Deoxyribonucleic acid is majorly found in the nucleus of cell. On the contrary, the RNA is found in both nucleus as well as in the cytoplasm. Cytoplasm is a semi-liquid substance which is made up with the help of the volume of the cell (Figure 4.6).

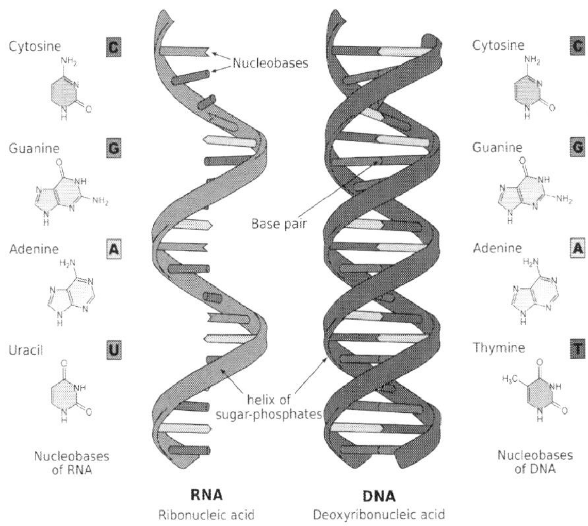

Figure 4.6. An illustration of molecular structure of DNA and RNA.

Source: Image by Wikimedia Commons.

Deoxyribonucleic acid and ribonucleic acid are different from one other with respect to their components. Deoxyribonucleic acid is consisting of the carbohydrate deoxyribose. On the other hand, ribonucleic acid has ribose. In addition to this, Deoxyribonucleic acid is consisting of base thymine, on the contrary, ribonucleic acid is consisting of uracil. The structure of DNA and ribonucleic acid is shown in Figure 4.7.

4.6. CLASSIFICATION OF ORGANIC COMPOUNDS

The organic compounds are classified into two main types as shown in Figure 4.7.

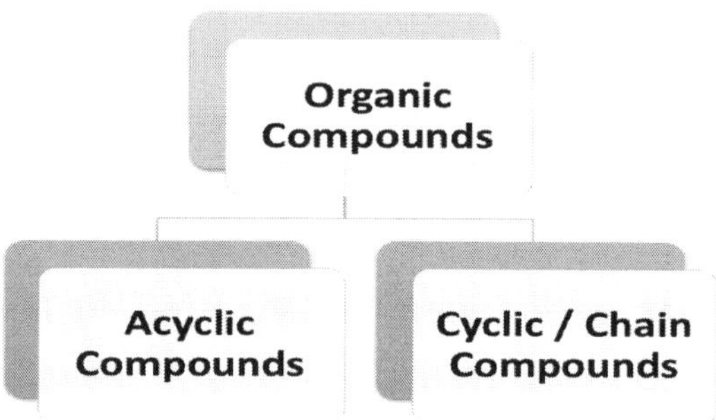

Figure 4.7. Classifications of organic compounds.

4.6.1. Open Chain or Acyclic Compounds or Aliphatic Compounds

The acyclic compound or open chain compounds are those compounds, in which the atom of the carbon is associated to each other. The association of the carbon either be in linear fashion or in branched fashion, in such a way that the molecules is having open chain structure. The open chain or acyclic compounds are also known as aliphatic compounds.

Examples are:

$CH_3\text{-}CH_2\text{-}CH_{3-}$ Propane

$CH_3\text{–}CH_{2-}CH_2\text{–}CH_3$ Butane

4.6.1.1. Acyclic Compounds

These are known as aliphatic compounds. They consist of straight chain or branched compounds.

In the field of chemistry, an open-chain compound or acyclic compound is a compound which is having a linear structure, in spite of having a cyclic structure. An open-chain compound having no side chains, is known as a straight-chain compound

Several numbers of the simple molecules of organic chemistry, for example, the alkanes and the alkenes, have both linear isomers as well as ring isomers, that is, both acyclic structure as well as cyclic structure, with the latter they are every so often categorized as aromatic. For those compounds which are having four carbon atoms or more than four carbon atoms, the linear forms can have straight-chain or branched-chain isomers.

The lowercase prefix n-shows the straight-chain isomer. For instance, n-butane is straight-chain butane, on the other hand, i-butane is isobutane. Cycloalkanes are the isomers that are made up with the help of alkenes, not of alkanes, because the closure of the ring is consisting of a C-C bond. Having no rings (aromatic or otherwise), all open-chain compounds are aliphatic.

Commonly, in the field of biochemistry, some of the isomers are more prevalent when they are compared to others. For instance, with respect to the living organisms, the open-chain isomer of glucose generally exists only transiently, in small amounts. Out of which D-glucose is the usual isomer, and, on the contrary, L-glucose is rare.

Straight-chain molecules are every so often not literally straight, in the context that their bond angles are often not 180°, but as the name of these compounds shows or depicts that they are schematically straight. For instance, the straight-chain alkanes are wavy or "puckered," as shown in the figure that is illustrated (Figure 4.8).

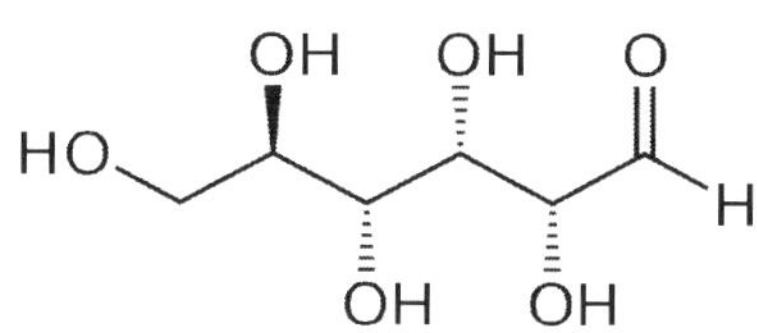

Figure 4.8. Open chain compound or acyclic compound: an illustration of open chain structure of glucose.

Source: Image by Wikimedia commons.

4.6.2. Closed Chain or Cyclic Compounds

Organic compounds with closed chain of atoms are called cyclic or ring compounds.

4.6.2.1. Cyclic Compounds

With respect to the field of chemistry, a cyclic compound (ring compound) is a term for a compound in which one series of atom or more than one series of atoms in the compound is linked in order to form a ring.

The size of the rings can fluctuate from three atoms to several numbers of atoms, and is consisting of an example, where all the atoms are carbon (that means, they are carbocycles). Out of which none of the atoms are carbon (that is known as, inorganic cyclic compounds), or where both carbon as well as non-carbon atoms are present (which is known as heterocyclic compounds).

According to the size of the ring, the bond order of the individual association or connection among the atoms of the ring, and their arrangements within the rings, carbocyclic compounds as well as the heterocyclic compounds can be aromatic or non-aromatic in nature. In the latter case, they can fluctuate from being completely saturated to having several numbers of multiple bonds between the atoms of the ring.

Due to the enormous variations allowed, in combination, with the help of the valences of common atoms and their ability in order to form rings, the number of possible cyclic structures, even of small size (for example,<17 total atoms) numbers in the many billions (Figure 4.9).

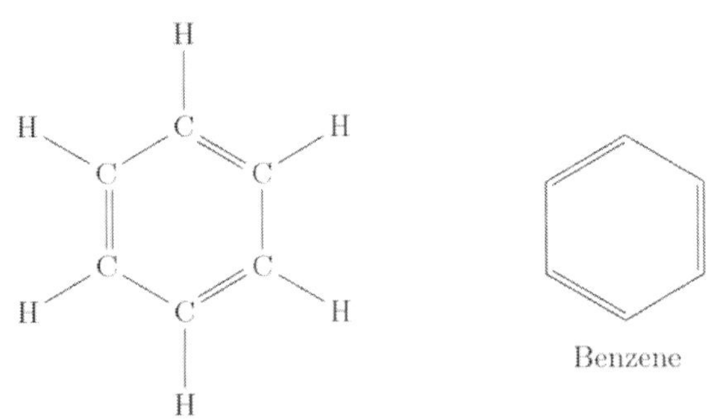

Figure 4.9. An illustration of cyclic compound: Molecular structure of benzene.

Source: Image by Latex Cook Book.

In addition to their complicatedness and the number, closing of atoms into rings can lock specific type of atoms with distinct substitution (with the help of functional groups) in such a way that stereochemistry as well as the chirality of the compound results, and also some manifestations that are distinct to the rings (e.g., configurational isomers).

Moreover, according to the size of the ring, the three-dimensional shapes of specific sort of cyclic structures, generally, they are rings which are having five atoms and more than five atoms, can fluctuate and interconvert in such a way that conformational isomerism is displayed. Certainly, the development of this vital chemical concept arose, historically, in reference to cyclic compounds.

In the end, the cyclic compounds, because of their distinctive shapes, reactivities, properties, as well as their bioactivities, becomes equally important as that of the straight chain ones. Cyclic compounds found to be the largest majority of all molecules that comes in the field of biochemistry, structure, and function of living organisms, and in the man-made molecules (e.g., drugs, herbicides, etc.) (Table 4.1).

4.6.3. Functional Group

A functional group may be defined as an atom or a group of atoms present in a molecule which largely determines the chemical properties

Table 4.1. Functional Group

Class of Organic Compounds	**Name of Functional Group**	**Structure**
Alkenes	Double Bond	-C = C-
Alkynes	Triple Bond	-C ≡ C-
Halogens	Halogen	-X (F, Cl, Br, I)
Alcohols	Hydroxyl	-OH
Aldehydes	Aldehydic (formyl)	-CHO

Carboxylic acids	Carboxyl	-COOH
Acid amides Primary amides	Amides Amino	$-CONH_2$ $-NH_2$
Ketones	Ketones	C = O

Source: Table from physicscatalyst.com.

4.6.4. Homologous Series

Homologous series is defined as a family or group of organic compounds that are having similar structure, and are repeated consecutively forming long chains of similar units of molecule in an organic compound. The length of the homologous series contributes to the size of a particular molecule.

4.7. HOMOCYCLIC COMPOUNDS

In these sorts of compounds, the structure of the ring is made up of only with the help of atoms of the carbon.

Homocyclic compounds are further classified into types of compounds that are mentioned below:

1. Aromatic compounds; and
2. Alicyclic compounds

4.7.1. Aromatic Compounds (Benzenoid)

Compounds that are having one ring of benzene or more than one benzene rings in their structure are known as aromatic benzenoid compounds. Most of the aromatic compounds have pleasant odor (Greek: Aroma-sweet smell).

Aromatic compounds are known to be a large class of unsaturated chemical compounds which are categorized by one planar rings of atoms or more than one planar ring of atoms connected or associated with the help of covalent bonds of two different kinds. The distinctive stability of aromatic compounds is referred to as aromaticity.

Nevertheless, the term aromatic originally related to the odor. In the present interval of time, the application of aromatic compounds in the field of chemistry is limited to compounds that have specific sort of electronic properties, structural properties, as well as chemical properties. Aromaticity is the consequence from specific kind of bonding arrangements that produce certain π (pi) electrons within a molecule in order to be held strongly.

Every so often, aromaticity is reflected in smaller than expected heats of combustion and hydrogenation and is linked with low reactivity.

Benzene (which is having molecular structure of C_6H_6) is the most popular aromatic compound. In addition to this, benzene is the parent to several numbers of other aromatic compounds are related. The six carbons of benzene are linked or associated in a ring, which are having the planar geometry of a regular hexagon in which all of the C—C bond distances are equal (Figure 4.10).

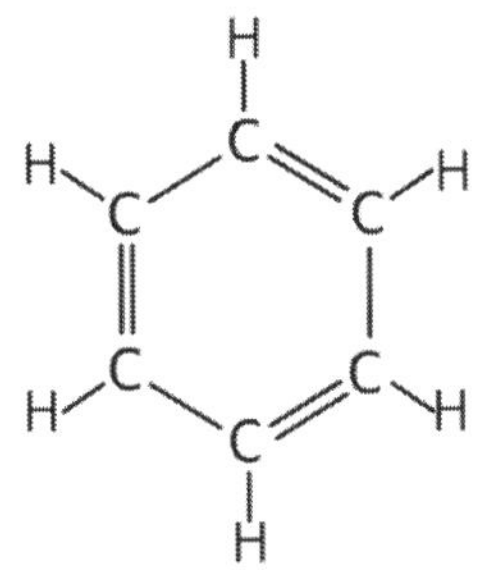

Figure 4.10. An illustration of aromatic compound: Benzene.

Source: Image by Wikimedia Commons

The six π electrons circulate in a region above and below the plane of the ring, each, and every electron being shared with the help of all six carbons. This step provides assistance in order to maximizes the force of attraction between the nuclei (which is positively charged) and the electrons (which are negatively charged).

In addition to this, equally important is the number of π electrons, which, depending on the molecular orbital theory, they must be equal to $4n + 2$, in which $n = 1, 2, 3$, etc. With respect to the benzene with six π electrons, $n = 1$.

The largest group of aromatic compounds are those in which, one hydrogens of the benzene or more than one of the hydrogens of benzene are substituted with the help of some other atom or group, as in toluene ($C_6H_5CH_3$) and benzoic acid ($C_6H_5CO_2H$). Polycyclic aromatic compounds are made up with the help of benzene rings that share a common side. For instance, naphthalene ($C_{10}H_8$).

Heterocyclic aromatic compounds are consisting of at least one atom other than carbon within the ring. Some instance that are including of pyridine (C_5H_5N), in which one atom of the nitrogen (N) substitutes the one CH group, and purine ($C_5H_4N_4$), in which two nitrogen replace by the two CH groups.

Heterocyclic aromatic compounds, such as furan (C_4H_4O), thiophene (C_4H_4S), and pyrrole (C_4H_4NH), which are consisting of five-membered rings in which oxygen (O), sulfur (S), and NH, respectively, replace an HC=CH unit.

4.7.2. Heterocyclic Compounds (Non-Benzenoid Aromatic)

Hetero cyclic compounds are those compounds, in which the ring atoms are made up with the help of hetero atoms such as atoms of nitrogen, atoms of oxygen and atoms of sulfur, in addition to atoms of the carbon are known as heterocyclic compounds.

Heterocyclic compounds are also known as heterocycles. Heterocyclic compounds are any of a major class of organic chemical compounds that are categorized with the help of the fact that some or all of the atoms in their molecules are linked in rings which is consisting of at least one atom of an element other apart from the carbon (C).

The cyclic part (from Greek "kyklos," meaning "circle") of the heterocyclic compounds shows that at least one ring structure is present in such a compound. On the other hand, the prefix hetero-(which is comes from Greek "heteros," meaning "other" or "different") suggests to the noncarbon atoms, or heteroatoms, that are present in the ring.

With respect to the general structure of the hereto cyclic compounds, the heterocyclic compounds resemble the cyclic organic compounds that include only carbon atoms in the rings. For instance, cyclopropane (which is having a three-carbon-atom ring) or benzene (which is having a six-carbon-atom ring).

But, on the contrary, the presence of the heteroatoms gives heterocyclic compounds physical properties as well as the chemical properties that are every so often quite different from those of their all-carbon-ring analogs (Figure 4.11).

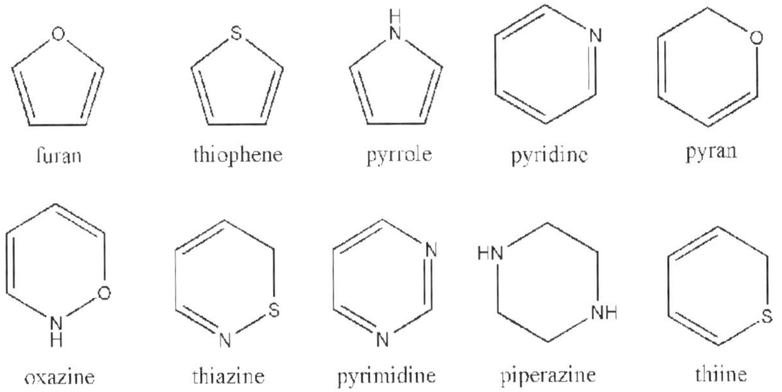

Figure 4.11. An illustration of heterocyclic compounds.

Source: Image by Wikimedia Commons.

Heterocyclic compounds are consisting of several numbers of the biochemical material that are very significant and important aspects of the life. For instance, nucleic acids, nucleic acid is the chemical substances that transport or convey the genetic information which is used to control or regulate the inheritance. Nucleic acid is consisting of long chains of heterocyclic units that are held together with the help of other sorts of materials.

Several numbers of naturally occurring pigments, vitamins, along with the various antibiotics are heterocyclic compounds in nature, as are most hallucinogens. In the present interval of time, the modern society is relying on synthetic heterocycles for the application of drugs, pesticides, dyes, along with the materials made of plastics.

4.7.3. General Aspects of Heterocyclic Compounds

The most common heterocycles are those, which are having five-membered rings or six-membered rings. In addition to that they are consisting of heteroatoms of nitrogen (N), oxygen (O), or sulfur (S). The best known of the simple heterocyclic compounds are pyridine, pyrrole, furan, and thiophene.

A molecule of pyridine is consisting of a ring which is having six atoms, out of which, five are atoms of carbon and one is nitrogen atom. Pyrrole, furan, along with the thiophene molecules each of them are consisting of five-membered rings, that is made up with the help of four atoms of carbon and one atom of nitrogen, oxygen, or sulfur, respectively.

Both pyridine as well as pyrrole are made up with the help of nitrogen heterocycles. The molecules of pyridine as well as pyrrole are consisting of atoms of nitrogen along with the atoms of carbon in the rings. The molecules of several numbers of biological materials are consisting in part of pyridine and pyrrole rings, and these sorts of materials yield small amounts of pyridine and pyrrole upon strong heating.

These substances were discovered during the 1850s in an oily mixture formed with the help of strong heating of bones. In the present interval of time, pyridine along with the pyrrole are made up with the help of or using the synthetic reactions.

The main commercial interest lies in their conversion to other substances, mainly the dyestuffs and drugs. Pyridine can also be taken into use as a solvent, a waterproofing agent, a rubber additive, an alcohol denaturant, and a dyeing adjunct.

Furan is a heterocyclic compound, which is consisting of oxygen atoms, that is employed mainly for conversion to other substances (including pyrrole). Furfural, which is a close chemical relative of furan. Furfural is obtained with the help of oat hulls and corncobs. Furfural is used in the production of intermediates for nylon.

Thiophene, which is a heterocycle composed with the help of atoms of sulfur, which resembles with the benzene with respect to its chemical properties as well as physical properties. Thiophene is a frequent contaminant of the benzene which is obtained with the help of natural sources and was first discovered through the course of the purification of benzene.

Similar to the other compounds, thiophene is mainly used for conversion to other substances. Both furan as well as thiophene were discovered in the latter part of the 19th century.

Generally speaking, the physical properties as well as the chemical properties of heterocyclic compounds are best understood when they are compared with the ordinary organic compounds, which are not having the heteroatoms.

4.7.4. Comparison with Carbocyclic Compounds

The molecules of the organic chemical compounds that are built up with the help of a framework or a long series of atoms of the carbon, in which the atoms of hydrogen (H), atoms of oxygen (O), and other hetero atoms are linked to it.

The atoms of the carbon have a distinct quality of being able to connect or associate with the other atom of carbon in order to create a chain of atoms. When the both of the ends of the chain are linked together in order to create or form a ring, as an outcome, the formation of cyclic compound takes place.

These kinds of substance are every so often are referred to as carbocyclic compounds. The substitution of one or more than one rings, which are made of carbon atoms in the molecules of a carbocyclic compound with a hetero atom gives a heterocyclic compound.

A generic example of carboxylic compound is cyclopentane (C_5H_{10}). The molecular structure of the cyclo-pentane is discussed Figure 4.12

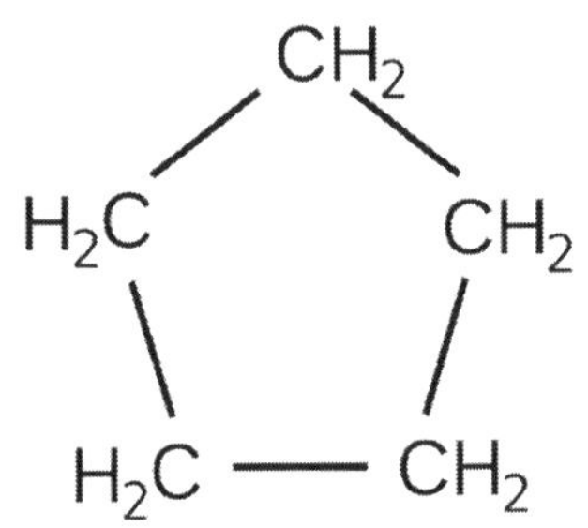

Figure 4.12. Molecular structure of cyclopentane.

Source: Image by Wikimedia Commons.

In the context of cyclopentane, the chemical symbols show that the atoms of the elements and the lines depicts the bonds between the atoms. For convenience these sorts of formulas, every so often they are written in the form of a simplified polygon.

With respect to the cyclopentane, in which each corner of the polygon depicts the presence of the atom of the carbon (it being understood that hydrogen is linked with the atoms of the carbon as per the requirement).

4.8. CONFORMATION OF ACYCLIC MOLECULES

Strain measures are the amount of energy, which is obtained by the molecule because of the structural distortion. Strain is the measure of the instability in the molecules which is generally an outcome of the repulsion of the electron.

All the molecules have an optional three-dimensional structure which permits for least possible strain in the system. On the other hand, several numbers of molecules alleviate this strain with the help of rotation of the bonds, while some of the molecules cannot.

The three-dimensional arrangement of the molecules which produces due to the rotation around any single bond is known as a conformation. It is possible to rotate around the C–C bond of ethane in order to determine the conformation (Figure 4.13).

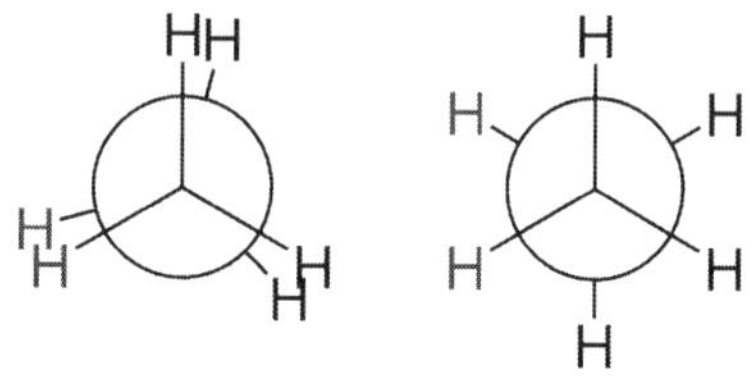

Figure 4.13. An illustration showing conformation of ethane.

Source: Image by Wikimedia Commons.

Talking about ethane, it can adopt a staggered conformation, as shown in the above-mentioned figure. This permits the hydrogen on either carbon to be as far as from each other as possible. Ethane can also adopt an eclipsed conformation, as shown in Figure 4.14.

In the ethane, the atoms of the hydrogen are as close as possible with respect to each other. Just similar to the solar eclipse, the atoms of the hydrogen in the front eclipse to those, that are present in the back side.

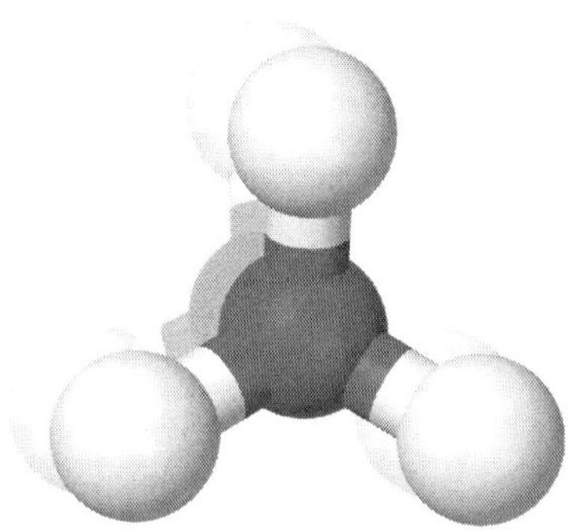

Figure 4.14. Eclipsed conformation of ethane from frond side to back side looking down the C-C bond.

Source: Image by Wikipedia.

A general term, which is used in explaining the elements of the conformation is dihedral angle. The definition of the dihedral angle (θ), is the angle between the two intersecting planes (a plane must me composed of at least three points).

In order to simplify, when explaining the dihedral angles, with respect to the conformation, it suggests to the angle which is present in between the two bonds. In the context of conformation, the angle between the two C-H bonds on each carbon is 0°.

The conformation that has the least strain will be the optimal arrangement. With respect to these two conformations of ethane, one is having the potential in order to decrease the strain much better, present in between the two atoms with the help of decreasing the repulsion of electron interactions of the hydrogen.

The staggered conformation is the lowest energy conformation, and the reason behind is that, the potential to get all the atoms as far from each other as possible. The eclipsed conformation is the highest energy conformation of the ethane. In this way, ethane will rotate about the C-C bond between these two conformations.

The difference in the energy of near about 12.6 KJ/mol. This is the measure of the torsional strain of the ethane, or the strain which can be alleviated with the help of rotating the bond (in the case of ethane, it is the carbon-carbon bond).

4.8.1. Butane

Now, talking about the structure of butane ($CH_3CH_2CH_2CH_3$). It should look something like this:

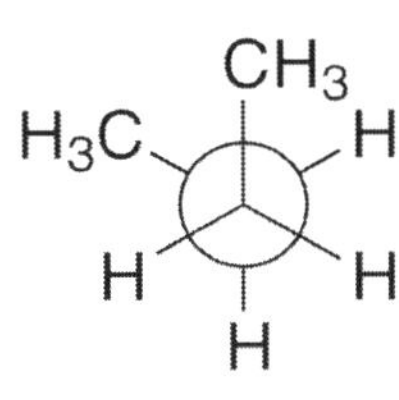

Figure 4.15. An illustration of conformation of butane.

Source: Image by Wikipedia.

In the context of butane, it has three carbon-carbon bonds. Since the rotation of the bond is permissible about all three, butane has more possible conformations when it is compared with the ethane. Now, explore the C2-C3 bond. When rotating around this bond, let's consider how to found the staggered and eclipsed conformations of ethane (Figure 4.15).

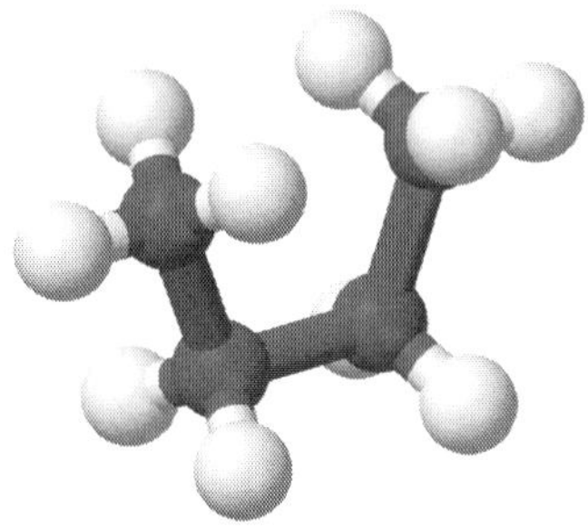

Figure 4.16. The methyl groups are eclipsed with hydrogens as seen when looking down C2-C3 and their dihedral angle becomes 120°.

Source: Image by Wikipedia.

Butane is consisting of two eclipsed conformations (as shown in the figure that is mentioned above). The groups which are taking participation in the 'eclipse' are different between the two conformations. If it is well thought out that the dihedral angles between the two C-CH_3 bonds, then the angles will be zero degrees and 120°, respectively for each of the conformation (Figure 4.16).

As, it rotates around C2-C3 between the two eclipsed conformations, it passes through gauche conformation of the butane, in which the two methyl groups have a dihedral angle of near about 60°. This is one of three staggered conformations of the butane.

Let's take a moment, in order to consider the relative strain of these conformers. Strain is linked to the size of an atom or a group. As cloud of the electrons gets bigger in size, so does the amount of repulsion its surroundings will experience.

Considering at the two eclipsed conformation, it can be seen that they have various amount of strain. With respect to the eclipsed conformation in which the dihedral angle is zero degrees, the methyl groups are the closest at the extent they can be, forcing the cloud of electrons in order to crowd each other.

When the angle of dihedral reaches to 120°, the methyl group is eclipsed with a hydrogen, and that is much smaller in size when it is compared with the methyl group. In this way, it is quite easy to assume that the closer the methyl with respect to each other, the higher the strain and the molecule will be less stable (or having much higher amount of energy) that conformation will be.

When comparing the gauche and the eclipsed conformation in which the dihedral angle is 120°, it can be noticed that the eclipsed conformation forces the methyl group in order to lie between the same plane, just similar to the hydrogen.

Since the cloud of the electron of the methyl group is much larger in size when it is compared with the single hydrogen, this conform is lower in energy, but it is still comparatively unstable, when it is compared to the other eclipsed conformer.

With respect to the gauche conformation, since all of the groups are staggered, the molecule is having the better potential in order to avoid the unnecessary strain. With respect to this, it can be concluded that the gauche conformation is lower in energy when it is compared with the both eclipsed conformers.

It the rotation of the methyl groups take place, in such a way that they are as far as possible and making an angle of dihedral of near about 180°, then anti conformation of butane could be determined.

This another staggered conformer of butane. Since all the groups that are present in the molecule are situated with as much space in between them as possible, this is the optimal, lowest energy conformation of butane.

With the help of this, a graph can be plot between the relative energies of these kinds of conformers as a function of their dihedral angles, as shown in the figure that is mentioned below. This will provide assistance in order to illustrate that the butane will exist predominantly in its anti and gauche conformation.

And the reason behind this is that, they have much higher stability because of the decrement in the torsional strain. And also, it is very significant to see that rotating from gauche to anti, it necessitates the molecule in order to pass through a conformer which is less stable, and which does not require any kind of external or internal energy (Figure 4.17).

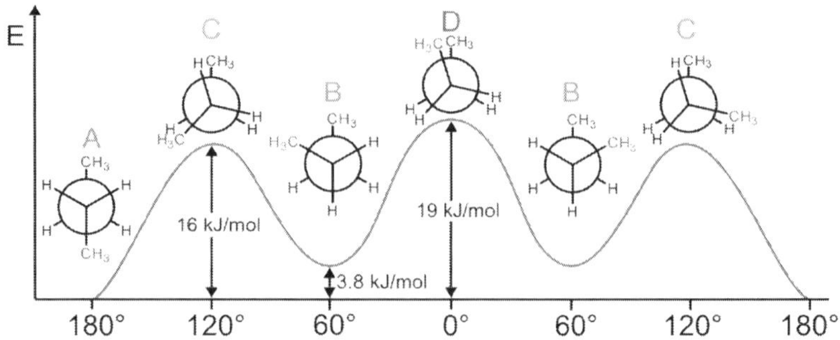

Figure 4.17. Plot of energy vs dihedral angle of butane's conformations.

Source: Wikimedia commons.

4.9. CONFORMERS OF CYCLOALKANES

Never the less, the customary line drawings of the simple cycloalkanes are geometrical polygons. In the matter of fact, the actual shape of these kinds of compounds in most of the cases are very dissimilar.

Cyclopropane is necessarily planar or flat in nature. Cyclopropane is consisting of the carbon atoms at each corner of an equilateral triangle. The bond angle is 60° which is much smaller when it is compared to the optimum angles of a normal tetrahedral carbon atom. The angle between the normal tetrahedral carbon atom is near about 109.5°.

As an outcome, the resulting angle strain dramatically impacts the chemical behavior of this cycloalkane. Cyclopropane also experiences substantial eclipsing strain, and the reason behind that is, all the carbon-carbon (C–C) bonds are completely eclipsed.

Cyclobutene decreases the some of the bond-eclipsing strain with the help of folding (the out of plane dihedral angle, which is near about 25°). But also, the total eclipsing and the angle strain remains high.

Cyclopentane has a very little angle strain (the angles of a pentagon are near about 108°), but its eclipsing strain would be large (near about 10 Kcal per mol), if it remained planar or flat. In this way, the rings which have five members, adopts a non-planar puckered conformation whenever it is possible.

The rings which are larger as compared to the cyclopentane, would have angle strain if they were planar. Nevertheless, this strain, altogether with the eclipsing strain inherent in a planar structure, can be relived with the help of puckering the ring.

Cyclohexane is a very good example of a carboxylic system, which virtually removes the eclipsing as well as the strain between the angles, with the help of adopting a non-planar conformation. Both cycloheptane as well as cyclooctane have greater strain when they are compared with the angle strain of cyclohexane, in large part, and the reason behind this is the transannular crowding (steric hindrance, which is caused by the groups on the opposite sides of the ring).

The cyclic systems are a bit different as compared to the open chain systems. With respect to the open chain system, any sort of bond can be rotated to 360°, going through several numbers of various conformations. That complete rotation is not possible in a cyclic system.

And the reason behind that is the parts that any individual would be trying to twist away, with respect to each other, would still be linked or associated together. Cyclic systems have fewer "degrees of freedom" as compared to the aliphatic systems. Talking about aliphatic systems, they have "restricted rotations."

Due to the fact of the restricted rotation of the cyclic systems, most of them have much more well-defined shapes, when they are compared to their aliphatic counterparts.

Let's take a look at the basic shapes of some common rings.

1. Several numbers of biologically important compounds that are built around the structures which are consisting of rings, in this way, it is very significant and important that scientists were become familiar with them.
2. With respect to the nature, three-membered rings to six-membered rings are frequently encountered.

4.9.1. Cyclopropane

A three membered ring has no rotational freedom whatsoever. A plane is defined with the help of three points, in this way, the three carbon atoms in cyclopropane are all constrained in order to lie in the same plane.

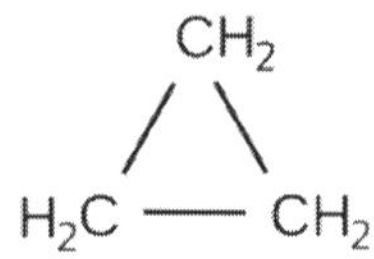

Figure 4.18. An illustration of cyclopropane.

Source: Image by Wikimedia commons.

In addition to this, if looking at a model, an individual will find that the neighboring C-H bonds (C-C bonds, too) are all held in eclipsed conformations (Figure 4.18).

Cyclopropane is always at maximum torsional strain. This strain can be illustrated in a line drawing of cyclopropane as shown from the side. In this oblique view, the dark lines mean that those sides of the ring are closer to you.

Nevertheless, the ring is not big enough in order to introduce any sort of steric strain, which does not become a factor until it reaches six membered rings. Until that point of time, rings are not flexible enough for two atoms in order to reach around and bump with respect to each other.

The really big challenging task with cyclopropane is that the C-C-C bond angles are all too small.

1. With respect to the cyclopropane, all the atoms of the carbon are found to be tetrahedral;
2. These bond angles ought to be 109°;
3. The angles in an equilateral triangle are actually 60°, which is about half as large as the optimum angle; and
4. This factor produces a huge amount of strain in the molecule, which is known as ring strain.

4.9.2. Cyclobutene

Cyclobutene is a four membered ring. In two dimensions, the shape of the cyclobutene is square, which is having 90° angles at each corner (Figure 4.19).

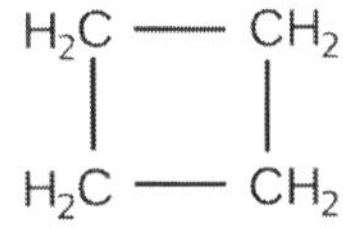

Figure 4.19. An illustration of cyclo-butane.

Source: Image by Wikimedia commons.

In the 3D form, cyclobutene is flexible enough, in order to buckle into the shape of a "butterfly," releasing torsional strain a little bit. When cyclobutene does that, the angles between the bonds get a little worse, and it decreases from the value of 90° to 88°.

In a line drawing, this butterfly shape is usually shown from the side, with the near edges drawn with the help of darker lines (Figure 4.20).

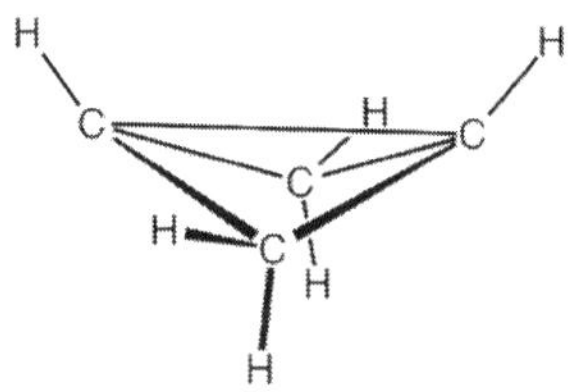

Figure 4.20. Different method to show the bonding in cyclo-butane.

Source: Image by Wikipedia.

1. With respect to the cyclobutene, it has the bond angles of 88° in spite of 109°. In addition to this, cyclobutene has a lot of ring strain, but less when it is compared with the cyclopropane;
2. In cyclobutene, the torsional strain is still existing, but the neighboring bonds are not exactly eclipsed in the butterfly;
3. Cyclobutene is still not large enough that the molecule can reach around in order to produce crowding. Steric strain is very low; and
4. Cyclobutene's are a little more stable when they are compared with the cyclopropane. In addition to this, cyclobutene's are also a little more common in nature.

4.9.2. Cyclopentane

Talking about the cyclopentanes, they are even more stable as compared with the cyclobutene. Cyclopentane is the second-most common paraffinic ring in nature, after cyclohexane. With respect to two dimensions, a cyclopentane appears to be a regular pentagon. With respect to three dimensions, there is enough freedom of rotation in order to allow a slight twist out of this planar shape. In a line drawing, this three-dimensional shape is drawn from an oblique view, just similar to the cyclobutene.

1. The ideal angle in a regular pentagon is near about 107°, which is very close to the bond angle of tetrahedral;
2. Cyclopentane distorts only very slightly into the shape of an "envelope," in which one corner of the pentagon is lifted up above the plane of the other four. In addition to this, as an outcome, the ring strain is completely removed;
3. The envelope eliminates the torsional strain along the sides and flap of the envelope. Never the less, the neighboring carbons are eclipsed along the "bottom" of the envelope, away from the flap. There is still some torsional strain in cyclopentane; and
4. Again, with respect to the cyclopentane, there is no steric strain exists in this system.

4.10. CONCLUSION

In the field of chemistry, the organic chemistry is the study of carbon along with the study of chemistry of life. In the time being, not all the reactions of the carbon are organic in nature. apart from that, another way to look at the organic chemistry would be well thought out about the study of the molecules that are having the bond between carbon and hydrogen (C–H), along with their reactions as well.

The study of the organic chemistry as well as the implementation of organic chemistry is very important in life. The chemical reactions that took place in the organic chemistry are also an important aspect of life of a human being along with the all living organisms.

Several careers apply the understanding of the organic chemistry, for example, doctors, veterinarians, dentists, pharmacologists, chemical engineers, and chemists. Organic chemistry or organic compounds plays a very significant part in the development of the common household chemicals, foods, plastics, drugs, and drives most of the chemical part of the daily life.

REFERENCES

1. (n.d.). Conformation of Acyclic Molecules. [ebook] pp. 1–10. Available at: http://www.chem.ucla.edu/~harding/tutorials/conformations_acyclic.pdf (accessed on 18 August 2020).
2. Alan, R. K., & Olga, V. D., (2020). *Heterocyclic Compound | Chemistry.* [online] Encyclopedia Britannica. Available at: https://www.britannica.com/science/heterocyclic-compound#ref277860 (accessed on 18 August 2020).
3. Brain Kart, (2020). *Classification of Organic Compounds: Open Chain, Cyclic Compounds.* [online] BrainKart. Available at: https://www.brainkart.com/article/Classification-of-Organic-Compounds--open-chain,-cyclic-compounds_2803/ (accessed on 18 August 2020).
4. Chemistry Libre Texts, (2019). *3.3: Conformations of Cyclic Organic Molecules.* [online] Chemistry Libre Texts. Available at: https://chem.libretexts.org/Bookshelves/Organic_Chemistry/Book%3A_Organic_Chemistry_with_a_Biological_Emphasis_v2.0_(Soderberg)/03%3A_Conformations_and_Stereochemistry/3.03%3A_Conformations_of_cyclic_organic_molecules (accessed on 18 August 2020).
5. Chemistry Libre Texts, (2019). *4.9: Conformations of Acyclic Alkanes—Ethane.* [online] Chemistry LibreTexts. Available at: https://chem.libretexts.org/Bookshelves/Organic_Chemistry/Map%3A_Organic_Chemistry_(Smith)/Chapter_04%3A_Alkanes/4.9%3A_Conformations_of_Acyclic_Alkanes%E2%80%94Ethane (accessed on 18 August 2020).
6. Cliffs notes, (2020). *Organic Compounds.* [online] Cliffsnotes.com. Available at: https://www.cliffsnotes.com/study-guides/biology/biology/the-chemical-basis-of-life/organic-compounds (accessed on 18 August 2020).
7. Dr. Dietmar, K., Prof. Steven, F., Chris, P. S., & William, R., (n.d.) *Conformers of Cycloalkanes | MCC Organic Chemistry.* [online] Courses.lumenlearning.com. Available at: https://courses.lumenlearning.com/suny-mcc-organicchemistry/chapter/conformers-of-cycloalkanes/ (accessed on 18 August 2020).
8. Francis, A. C., (2020). *Aromatic Compound | Chemical Compound.* [online] Encyclopedia Britannica. Available at: https://www.britannica.com/science/aromatic-compound (accessed on 18 August 2020).

9. Olga, V. D., & Alan, R. K., (2020). *Heterocyclic Compound - Comparison with Carbocyclic Compounds.* [online] Encyclopedia Britannica. Available at: https://www.britannica.com/science/heterocyclic-compound/Comparison-with-carbocyclic-compounds (accessed on 18 August 2020).
10. Olga, V. D., & Alan, R. K., (2020). *Heterocyclic Compound - The Nature of Heteroaromaticity*. [online] Encyclopedia Britannica. Available at: https://www.britannica.com/science/heterocyclic-compound/The-nature-of-heteroaromaticity (accessed on 18 August 2020).
11. Physics catalyst, (2019). *Classification of Organic Compounds: Cyclic and Acyclic Compounds*. [online] Physicscatalyst.com. Available at: https://physicscatalyst.com/chemistry/classification-of-organic-compounds.php (accessed on 18 August 2020).
12. Science direct, (2020). *Open-Chain Compound - An Overview | Science Direct Topics*. [online] Sciencedirect.com. Available at: https://www.sciencedirect.com/topics/pharmacology-toxicology-and-pharmaceutical-science/open-chain-compound (accessed on 18 August 2020).
13. The Editors of Encyclopedia Britannica, (2019). *Hyaluronidase | Enzyme*. [online] Encyclopedia Britannica. Available at: https://www.britannica.com/science/hyaluronidase (accessed on 18 August 2020).
14. Wikipedia, (2019). *Cyclic Compound.* [online] En.wikipedia.org. Available at: https://en.wikipedia.org/wiki/Cyclic_compound (accessed on 18 August 2020).
15. Wikipedia, (2019). *Open-Chain Compound*. [online] En.wikipedia.org. Available at: https://en.wikipedia.org/wiki/Open-chain_compound (accessed on 18 August 2020).

Chapter 5 Symmetry and Configuration of Organic Compounds

CONTENTS

The symmetry and configuration of the organic compounds have been evaluated in this chapter as on the basis of these two factors the chemical structures are dependent. Both, symmetry, and configurations will be simultaneously described. At first the factors in relation to symmetry of organic molecules have been evaluated that includes elements and operators of symmetry.

Classification of molecules is performed considering the symmetry of the molecules, this has been also evaluated later in the chapter. In terms of configuration, the two major factors related to the configuration that are relative and absolute configuration have been described efficiently. Thus, this chapter provides all the relevant data related to the symmetry and configuration of organic compounds.

5.1. INTRODUCTION

The essential aspects of the organic compounds that are symmetry and configuration have been described here by considering every important factor in relation to stereochemistry. The impact of these aspects is evaluated by considering their types and different working frameworks.

Symmetry is a quality of items found in design and in different types of art just as in the domain of nature. It additionally has a significant impact in science: in Molecular spectroscopy, in quantum mechanics just as (in the current situation) in the assurance of structure and the comprehension of stereochemistry.

The constituent of symmetry is the regular recurrence of specific examples inside an object or structure (Orchin and Jaffe, 1970, 1971; Donaldson and Ross, 1972; Scenternikov and Koptsik, 1974; Kettle, 1985; Hargittai and Hargittai, 1987; Heilbronner and Dunitz, 1993). As called attention to, it is frequently helpful to consider atom's static substances that can be considered by unbending mechanical models.

The symmetry connections to be considered refer to such perfect atoms or Molecular models. One should remember that, since atoms are three-dimensional (3D), generally just 3D portrayals will be totally satisfactory as models.

In this manner the tetrahedron representing (+)-lactic corrosive or its drawing is a legitimate model, yet the Fischer projection (without the fitting determinations regarding which groups are in front and which are in the back) is not.

An amateur thinking about the formulas of (+)-and (–)-lactic acids in their Fischer projections may reach the mistaken resolution that these structures have a plane of symmetry and are superposable (as they would be if carbon were square planar instead of tetrahedral).

Configuration has been characterized (Cross and Klyne, 1976; IUPAC, 1979) as "The course of action of the atoms in space of an atom of characterized constitution regardless of plans that vary just as after axis around at least one single bonds."

According to Noel (1987) the term configuration was presented by Wunderlich in 1886 (Auwers and Meyer, 1888). Victor Meyer, one of the pioneers of stereochemistry, has characterized it as the "geometrical" part of structure.

The IUPAC definition presents a few challenges and is, actually, not all around concurred on for reasons. One trouble, is that bond request may change persistently, and it is inconvenient, if certainly feasible, to single out revolution about a solitary bond.

Are the C-N bonds in the amides and thioamides to be viewed as single or fractional double bonds, and are the cis and trans isomers of these mixes consequently to be considered as contrasting in configuration, or would they say they are essentially conformational isomers of mixes of exceptional configuration?

A similar issue emerges in biphenyls and in various different structures. Similarly, with the 2-bromobutanes the issue is identified with the procedure of perception.

Physical scientific experts, whose system (e.g., vibrational spectroscopy) is equipped for identifying species isolated by very low boundaries would in general characterize configuration without the exception in regards to axis about single bonds.

Configuration just refers to the spatial game plans of bits in an atom of given structure, that is, to the stereochemical part of structure, as initially visualized by Victor Meyer who, accepted revolution about single bonds to be free.

The different types of butane at that point become configurational isomers. Not many organic scientific experts utilize the term configuration right now.

On the contrary, it has been recommended that, in the meaning of configuration, no respect to be given to turn about responsibilities of any request, including double bonds. Under this definition, cis-and trans-2-butene would be conformers, not configurational isomers; it is justifiable that such a view has not become mainstream either.

It is perceived that this definition is very flawed for a few reasons, one being the referenced vulnerability with respect to what establishes a solitary bond.

Another fact is that it discretionarily takes segregation (as opposed to some other factor, for example, spectroscopic discernibleness by one or other system) as the foundation of presence of configurational isomers, that is, it manages "remaining isomers" utilizing the paradigm of isolability.

An increasingly genuine defect is that "isolability" is itself not a clearly characterized term, in that it needs reference to both the temperature of the trial and the lifetime required to consider a species group "isolable." Nonetheless, the measure has merit from a thermodynamic perspective since, in use of the stage rule, just isolable species consider separate parts.

5.2. SYMMETRY ELEMENTS

The symmetry of an atom is considered by the presence of symmetry activities performed as for symmetry components. A symmetry component is a line, a plane or a point in or through an object, about which a revolution or reflection leaves the item in a direction undefined from the first.

A plane of symmetry is assigned by the image σ (or sometimes s), and the reflection activity is the casual event of atoms on one side of the plane with relating molecules on the opposite side, just as reflected in a mirror.

An inside or purpose of symmetry is named I, and the reversal activity shows unexpected event of every atom with an indistinguishable one on a line going through and an equivalent way from the reversal point (chair cyclohexane) (Figure 5.1).

Figure 5.1. Symmetry element: Chair cyclohexane.

Source: Image by Wikimedia.

At last, a rotational axis is assigned C_n, where the degrees of revolution that reestablish the object is 360/n (C_2= 180° turn, C_3= 120° revolution, C_4= 90° pivot, C_5= 72° pivot). C_1 is known as the character activity E since it restores the first direction.

An object having no symmetry components other than E is called uneven. Such an object is essentially chiral. Since a plane or purpose of symmetry includes a reflection activity, the nearness of such a component makes an object achiral. At least one rotational axes of symmetry may exist in both chiral, dissymmetric, and achiral objects.

One more symmetry activity must be characterized. Both trans-dimethyl cyclopropane and 1,3,5,7-tetrafluoro–1,3,5,7-cyclooctatetraene have a C_2 center, and both do not have a plane or focal point of symmetry.

The previous is chiral, yet the last is achiral in light of the fact that it has a S_4 inappropriate rotational pivot (also called an exchanging center).

An inappropriate center, S_n, comprises of a n-overlay revolution finished by reflection a mirror plane opposite to the turn pivot (n is constantly 3 or bigger in light of the fact that $S_1 = \sigma$ and $S_2 = I$). This is proportional to stating that a n-crease pivot changes over an object into its perfect representation.

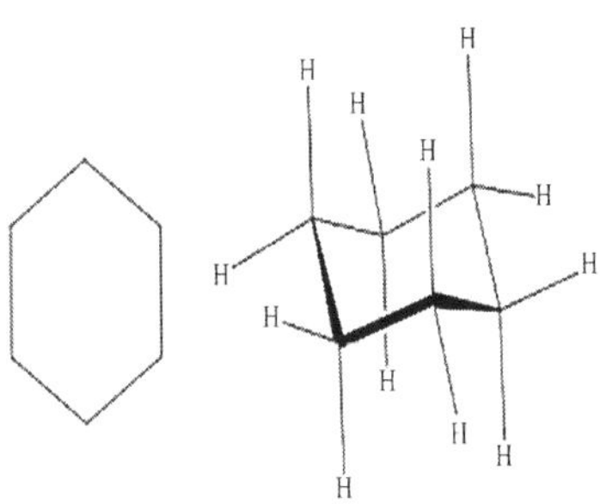

Figure 5.2. Chair and skeletal structures of cyclohexane.

Source: Image by Flickr.

Symmetry components are the administrators that produce the recurrent example of symmetry. In limited objects they are the appropriate pivot of symmetry C_n, the plane of symmetry, the focal point of symmetry I, and the exchanging or inappropriate center of symmetry S, (Orchin and Jaffe, 1970a) (Figure 5.2).

In objects of interminable degree (on the atomic scale, for example, gems, and polymer objects, there is an extra component of translational symmetry which, related to reflection, may offer ascent to float planes and, related to turn, to screw axes (Donaldson and Ross, 1972).

The center of symmetry, of variety n, additionally called the "n-overlap center" and indicated by the image C, is a center to such an extent that in the event that one turns the model (or atom) around the center by 360°, the new situation of the model is superposable with the first one.

Models are cis-(lR, 3R)-sec-butyl cyclobutene, which has a double (C_2) axis and r–1, c–2, c–3, c-4-(1R, 2R, 3R, 4R)-tetra-sec-butyl cyclobutene, which has a fourfold center of symmetry (C_4). With respect to, c represents cis, t for trans, and r for reference gathering.

It is evident that any figure or model, turned 360° around any center will be superposable upon itself. The C_4 symmetry pivot is an all-inclusive (and thus insignificant) symmetry component; it is equal to the "character activity," (E or I) in group hypothesis.

Turn is a "genuine" activity as in the emphases that are brought into superposition are real, material focuses. Such an activity is known as a "symmetry activity of the principal kind" or "appropriate activity" (Kettle, 1985).

The proximity of a symmetry center doesn't block chirality; in reality, both of the atoms are chiral. Consequently, chirality can't be compared with asymmetry (i.e., the absolute nonappearance of symmetry); in the more established form "dissymmetry" is frequently utilized as an equivalent word for what is presently call chirality.

Planes, focuses, and substituting axes of symmetry are components relating to "symmetry tasks of the subsequent kind" as per the fact that, opposed to bringing into unexpected event material focuses, they carry into one material point with the impression of another, a coincidence that may be named virtual instead of genuine.

A plane of symmetry will be a reflection plane by the activity of which each piece of a model (every atom of an object) is carried into unexpected event with a like part (or molecule), found somewhere else in the model (or atom). The compound cis-(1R,3S)-di-sec-butyl cyclobutene gives a model; the plane demonstrated is a plane of symmetry.

The central point of symmetry I is a point with the end goal that if a line is drawn from any piece of the model (or atom in a molecule) to that point and expanded an equivalent separation on the opposite side, a like part (or atom) is experienced.

A case of an atom with a focal point of symmetry is trans-(lR,3S)-di-sec-butyl cyclobutene: The focal point of the four-membered ring is the focal point of symmetry, some of the time additionally called "purpose of reversal." If the purpose of reversal is taken as the origin of a Cartesian facilitate framework, the impact of the reversal activity is to transpose a state of coordinates x, y, z into a state of coordinates-x,-y,-z.

5.3. SYMMETRY OPERATORS

A symmetry activity is an activity that conveys an atom (or a model or any figure or set of focuses) into a position unclear from (or proportional to) the first position. (Mislow, 1965; Salthouse and Ware, 1972; Orchin and Jaffk, 1970b.)

It is realized by the activity of one or a few of the prior referenced symmetry components that along these lines work as "symmetry administrators." For instance, progressive activity of the C_4 axis in the atom, B beginning with the first atom, offers ascend to three extra, superposable species turned around the C_4 center by 90," 180°, and 270°, individually; revolution by 360" drives back to the first position. These symmetry tasks so performed are called personality activity.

Chiral atoms essentially have a place with point groups that is, groups that have just appropriate axes. All other point groups, that is, those containing substituting axes, including reflection planes and reversal focuses, are related with achiral atoms.

5.3.1. Symmetry in Simple Carbon Compounds

Molecular geometry is identified with Molecular symmetry: this might be outlined by taking a look at some basic organic mixes. A quadrivalent carbon atom might be connected rightfully to one, and at generally four, different atom or nuclear grouping, for the most part called ligand.

The quantity of ligands in this manner connected to a focal molecule is named the coordination number of the last mentioned. Just two one-facilitated carbon combinations are known, carbon monoxide and the cyanide anion.

Both contain an axis and are of cone shaped symmetry. Every one of the two-arrange carbon mix are straight, their symmetry being either funnel shaped (e.g., HCN) or round and hollow [e.g., acetylene or carbon dioxide).

In the last, the Molecular axis is cut up by a plane of symmetry. Three-arrange mixes of carbon are normal; accordingly, the cross-breed carbon atom is constantly 3-facilitating, its related ligands being appropriated in a planar manner.

Three indistinguishable ligands around a sp^2-hybridized carbon atom in a steady compound just in the carbonate anion or in the graphite cross section. In temperamental elements, the sp^2-hybridized carbon molecule is similarly 3-facilitating as in carbonium atoms, carbanions, and carbon radicals.

The geometry of the carbonate atom. It has eight symmetry components; one C_3 axis, three C_2 axes, and four symmetry planes. In the event that one of the ligands is not quite the same as the others (as in formaldehyde) the quantity of symmetry components is decreased to three: one C_2 axis and two c-planes.

This reducing in symmetry is not just of hypothetical logic however has impact the real elements of an atom. Since various interesting and shocking collaborations happen between various sets of atoms, it follows that the interatomic separates in the subsequent atoms will similarly be unique.

These distinctions in bond lengths and edges are normally quantifiable and hence the bond point of 120° expected for the sp^2-hybridized carbon molecule is just a guess, in light of the fact that the genuine estimation of this edge will rely upon the ligands related with the carbon atom.

At the point when three unique ligands are included just the cr-plane which concurs with the plane of the atoms is held, and it follows, consequently, that every one of the three bond lengths and edges must be unique.

One of the essential fundamentals of organic auxiliary hypothesis is that the ligands related with a 4-arrange carbon atom can't be coplanar yet are set at the apices of a tetrahedron. At the point when every one of the four ligands are indistinguishable this tetrahedron is standard with bond edges all equivalent to 109.5°.

Four-arrange carbon structures having four indistinguishable ligands (e.g., methane, carbon tetrachloride, neopentane) are very normal. Such mixes, as encapsulated via carbon tetrachloride, have four C_3 axes which concur with the four C–CI bonds and three C_2 axes which separate the CI–C–CI security points.

Every CI–C–CI set of three additionally characterizes one plane of symmetry, and since there are six unique manners by which any two ligands might be consolidated out of four it follows that there are six planes.

5.4. CLASSIFICATION OF MOLECULES ACCORDING TO THEIR SYMMETRY

Objects can be characterized by the nature and number of their symmetry components. Certain symmetry components in a given molecule, for example, the G_z axis in ethane which agrees with the C–C bond, might be invariant to conformational change, for example, to turn about single bonds, while different components just emerge in specific adaptations.

For a non-rigid molecule symmetry must be determined for a distinct compliance. In practice, it is common to pick a specific adaptation which might be viewed as a period normal of the infinite potential compliances which may emerge.

The characterization of molecules might be enormously disentangled through the presentation of a third, composite symmetry component, the turn reflection axis which is for the most part meant as 8n.

The symmetry activity performed by a Sn axis, for example, by a revolution reflection axis of the nth request, interfaces indistinguishable focuses through turn by 360° and by reflection through a plane opposite to S_n.

By necessity, the request for the axis is even or 1. It very well may be comprehended that a S axis is equal to a mirror plane, since right now is performed without axis. The S_2 axis is otherwise called a focal point of symmetry or reversal focus (G); on any straight line going through this inside indistinguishable component might be found at focuses equidistant from the middle.

From a symmetry perspective, objects (or atomic models) might be either chiral or achiral. A chiral object cannot be superimposed on to its identical representation, though with achiral objects such superimposition is conceivable through the presentation of specific interpretations as well as rotations."

It ought to be called attention to, however, that optical movement may just be seen in compounds which are chiral in the entirety of their moderately steady compliances. Chiral atomic models need Sn axes, while the nearness of Gn axes is perfect with chirality.

At the end, optically-dynamic molecules are not really unbalanced. Molecular models which are chiral or achiral might be additionally subdivided by their symmetry into supposed point groups, for which an arrangement of documentation was proposed by Schonflies (Figure 5.3).

Figure 5.3. Example of point group.

Source: Image by Wikipedia.

Molecular asymmetry is not limited to models containing focuses of asymmetry; a get together of molecules without a hazzle may display general asymmetry. Both the allene and the biphenyl are asymmetrical.

In the last case, substituents in the ortho-position keep the molecule from taking up an achiral symmetrical compliance. Chirality is good with basic axes of symmetry. Therefore, molecules having a solitary axis of symmetry have a place with the point group C_n (C_2, C_3, and so on.).

Compounds of C_2 symmetry are very normal. Such structures can be superimposed on to the first after pivot through 180°, yet not after reflection through a mirror plane. Invertible (non-coordinated) tube shaped helices with a uniform pitch are chiral and have C_2 symmetry. Helicenes, for example, heptahelicene, are customary invertible atomic helices.

5.5. MOLECULAR CLASSIFICATION BY THEIR SYMMETRY

The connections between at least two molecules will be analyzed, the idea of isomerism suggesting that examination must be made between two elements. The order of molecules based on symmetry standards alone is very straightforward.

At the point when two models are analyzed, the principal premise ought to be whether they are super imposable or not. In the event that the models are indistinguishable there is no compelling reason to inspect them further.

However, two non-indistinguishable models with a similar constitution might be connected as object and perfect representation or something else. Non-indistinguishable models (stereoisomers) which are connected as identical representations are called enantiomers.

Different sets of stereoisomeric models are called diastereomers. Since a given object can't have more than one identical representation, it is difficult to have more than one enantiomer of a given model. The quantity of conceivable diastereomerically related sets relies upon the constitution of the molecule and may now and again, for instance with steroids, be in excess of a hundred.

Since isomerism consistently suggests a distinction, just such models can offer ascent to sets of enantiomers whose perfect representations are not indistinguishable with the first. Such a property is displayed by models which have a place with the chiral point gatherings, for example, to the point groups Cn and Dn.

The two enantiomers of a couple fundamentally fall into a similar point group; diastereomers could possibly fit in with this conduct as a general rule they have various balances. Different enantiomeric and diastereomeric connections might be exhibited by considering the epoxy but-2-enes. The cis-epoxide (C), got from (Z)-But-2-ene-1, has a mirror plane; it is in this manner achiral and indistinguishable with its perfect representation (Figure 5.4).

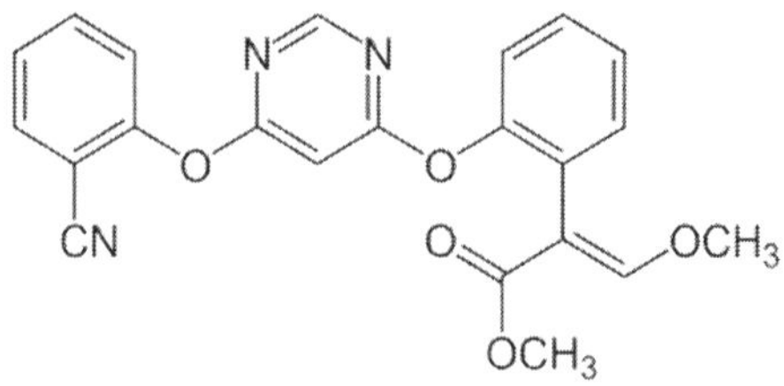

Figure 5.4. Structure of epoxy but-2-enes.

Source: Image by Wikipedia.

Having set up that a couple of models is stereoisomeric, the following undertaking is to characterize the idea of the distinction between them. With diastereomers this is a moderately clear undertaking and can generally be decreased to contrasts in interatomic separations, for example, to scalar geometrical parameters.

In this way, for instance, it may conclude that on account of the 2-enes one ought to be named Z (cis) since the separation between the terminal carbon atoms is littler than in its pair, the E (trans) diastereomer.

Consequently, diastereomers may contrast in the entirety of their scalar properties including reactivity, in spite of the fact that this last distinction may not be perceptible tentatively. For sets of enantiomers it follows that the relating geometrical parameters should all be equivalent.

Hence enantiomers can't be separated as far as geometry however just as far as a course, for example, as far as a vector as opposed to a scalar amount. Along these lines the trans-epoxibutanes (Tr) and (Tr*) might be recognized as far as the bearing where the hydrogen on the epoxide ring focuses towards the vicinal methyl gathering.

For (Tr*) this course is clockwise, for (Tr) anticlockwise. To recognize the two enantiomers, we have been compelled to receive a show, which right now identified with the development of the hands of a clock.

In this manner one can relate the chirality of the model being talked about to the chirality of another chiral object, an alleged chiral standard. In science the standard received is the left or right human hand.

The strategy utilized above to recognize the enantiomers of Jraws-epoxibutane is of no compound worth, since practically molecules cannot be straightforwardly watched. From the exploratory perspective a basic physical property which might be utilized to separate between enantiomers is optical axis (Figure 5.5).

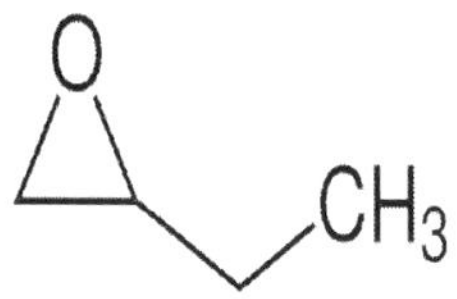

Figure 5.5. Epoxibutanes structure.

Source: Image by Wikipedia.

The outright estimation of the revolution of enantiomers is equivalent, and tragically there is no immediate connection between the indication of the optical turn and the special course of action of the atoms in chiral molecules.

Over 100 years passed after the revelation of optical action before such a connection could be first settled. Enantiomers might be distinguished, however, without falling back on optical pivot if there is some manner by which an association between two perfect representation enantiomer molecules and two molecules of a similar sort of a similar chirality can be set up.

Two right hands fit obviously superior to one side and a left. The blend of enantiomers gives a totally unique outcome relying upon whether such mix is with an indistinguishable or an identical representation species category.

In this way, two sorts of lactides are considered: that shaped from a similar enantiomer of lactic corrosive (LA) + (LA) and that from antipodes (LA) + (LA*). The subsequent lactides (LA–LA and LA–LA*) are not enantiomers but rather diastereomers.

While the methyl gatherings, for instance, are on a similar side of the ring in (LA–LA) they are on inverse sides in (LA–LA*). Right now, scalar physical properties of the items, for example, their dissolving focuses are extraordinary, just like the total estimations of their optical pivots.

As opposed to utilizing a stereochemical show, the enantiomers of lactic corrosive can be recognized through scalar physical properties. This outcome is as per the general rule that the collaboration of a chiral species categories with the antipodal individuals from any enantiomeric pair (even with its own sort) is a diastereomeric procedure, for example, it is diverse from a scalar perspective.

Indeed, even the various headings of optical turn saw with enantiomers might be followed back to diastereomeric communications.

5.6. ABSOLUTE CONFIGURATION

The exact or detailed arrangement of the substituents at a stereogenic center is known as the absolute configuration of that particular molecule. If an individual has to draw the structure, then he or she can use the R-and S-notation system in order to locate the absolute configuration of that structure of that particular molecule.

It is a more considerably a difficult task in order to be able to locate the absolute configuration to an actual sample of a molecule (that is a compound in a bottle).

This process is can be achieved with the help of solving the x-ray crystal structure of a molecule (which is a method that is not always readily available), with the help of spectroscopic methods, or possibly with the help of inference based on chemical reactions of known specific stereochemistry involving a compound whose absolute configuration is known.

The absolute configuration and the optical rotation of the starting material are known, and the configurations of the each of the subsequent products can be inferred with the help of knowing the stereochemistry of the reactions that take participation, the meaning of the optical rotation cannot be easily anticipated and it required to be identified with the help of an experimental measurement.

Similarly, as established isomers are portrayed as equations, it is essential to give separating images (supposed descriptors) to enantiomers, for example, those of the lactic acids appeared, in three-dimensional (3D) or projection formulas.

Obviously, one can portray the two acids tentatively by the detectable indications of their revolution however one likewise needs to determine their 3D structure without falling back on an equation drawing. It is with the end goal of such determination that frameworks of configurational documentation have been conceived.

The generally utilized Cahn-Ingold-Prelog (CIP) framework (Cahn et al., 1956, 1966; Prelog and Helmchen, 1982) for the determination of atomic configuration started around the same time (Cahn and Ingold, 1951) in which outright configuration was first tentatively controlled by Bijvoet (Figure 5.6).

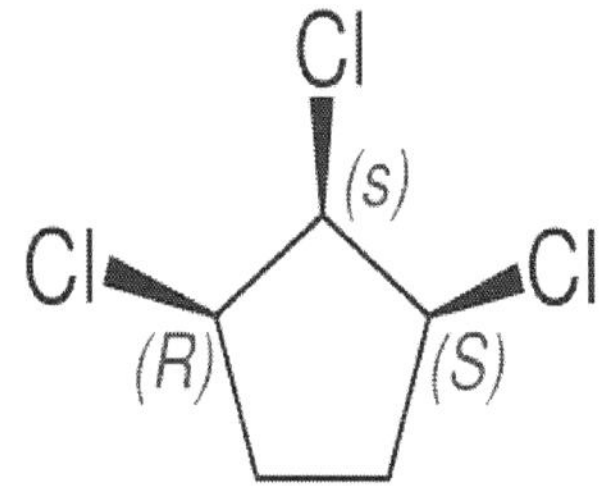

Figure 5.6. Cahn-ingold-prelog (CIP) framework.

Source: Image by Wikipedia.

Right now, (emphasized) descriptors R and S, which are set in enclosures and associated with a hyphen just when utilized as prefixes (configurational descriptors) of compound names, are utilized to indicate the configurations of enantiomers and diastereomers.

To decide the correct descriptor for a chiral focus, the CIP chirality rule (Cahn, Ingold, and Prelog, 1966) is utilized. The four ligands connected to the chiral focus can be masterminded in an interesting arrangement; for the minute let us consider these ligands A, B, D, and E with the stipulation that the succession (or need) is A goes before B goes before D goes before E (A > B > D >E or A+ B+ D+ E).

The CIP chirality decide necessitates that the model be seen from the side inverse to that involved by the ligand E of least need. The staying three ligands at that point present a tripodal cluster, with the legs stretching out toward the watcher.

If the consecutive course of action or ability to read a compass of these three ligands (A-, B-, D) is then clockwise, the configurational descriptor is R, in the event that it is counterclockwise, it is S.

At the point when at least two atoms legitimately joined to the chiral focus are the equivalent, as on account of lactic corrosive, a guideline of outward investigation is applied: If the proximal molecule appended to the chiral focus does not give a choice, one continues apparently to the subsequent atom, at that point, if necessary, to the third, etc.

On account of the lactic acids, since the carbon molecules of CO_2H and CH, give no choice, one continues to the following connected atom and notes that 0 > H. The correct execution of this tree-chart investigation requires an all-around characterized order (Prelog and Helmchen, 1982).

All ligands in a given circle must be investigated before one continues to the following circle. In this way in P-methoxy lactic corrosive, after one discovers one oxygen molecule in the subsequent circle connected to every carbon, one must not continue to the third circle, yet one must investigate the subsequent circle further and note that there is a second oxygen atom joined to the carbon in CO_2H yet not in CH_2OCH, and since O >H, CO_2H> CH_2OCH.

Interestingly, however, when a priority of one way of investigation over another has been set up in one circle, that priority conveys to the following circle.

Assume one needs to decide the grouping of the two ligands. Unmistakably, no choice is reached at the main (C), second (C, C, H), or third (C, H, F, C, H, Br) circle. One hence needs to continue to the fourth circle, and one does utilize the way of higher priority at the third circle (Br > F).

The choice, that ligand B goes before ligand A, is made on the premise that Br > C1 in the fourth circle of the favored branch as opposed to base on the perception that (I) is the most elevated nuclear number atom in the fourth circle (of the less considered branch).

5.6.1. Determination of Absolute Configuration

1. Bijvoet Method

It was distinctly in 1951 that the primary declaration of the total configuration of any chiral molecule was accomplished (Bijvoet et al., 1951; Bijvoet, 1955). In common X-beam crystallography the forces of the diffracted shafts rely upon the separations between the atoms yet not without a doubt the spatial direction of the structure.

This comes about on the grounds that the stage change because of dissipating of the occurrence radiation is (almost) the equivalent for all molecules. A chiral crystal and its enantiomorph produce a similar X-beam designs and cannot be recognized from each other.

As a basic model, the one-dimensional (ID) chiral exhibit A-B can't be recognized from B-An in light of the fact that the impedance design anytime I in space will be the equivalent for the two 1D enantiomers since it relies just upon the total estimation of the length contrast IAI-BI1 = 1BI-A11.

Consequently, there will be no adjustment in impedance design when atoms A and B are exchanged. The equivalent is valid in the more perplexing instance of a 3D chiral precious stone: Since there is nothing to recognize engendering of X-beams toward a path A-B from that the other way B-A (Friedel's law), X-beam diffraction designs are centrosymmetric, regardless of whether the crystal under scrutiny has a focal point of symmetry or not (Dunitz, 1979).

The answer for this issue was, on a basic level, found in 1930 when Coster et al. (q. v.) utilized alleged abnormal X-beam dissipating to decide the succession of planes of zinc and sulfur atoms in a crystal of zinc blende (ZnS).

The strategy includes utilizing X-beams of a frequency close to the ingestion edge of one of the atoms, right now. This outcomes in a little stage change of the X-beams dispersed by the zinc atoms comparative with the sulfur atoms, which does not rely upon their relative positions.

If B indicates the zinc atom, the impact of hindering the period of the diffracted X-beam is equivalent to if the way B-I were longer than an I, by a speculative additional way length p, recreated by the stage change.

Hence the dispersed forces for A-B and B-An are not, at this point the equivalent, since the stage change is in inverse ways in the two cases. Regarding the speculative amount p, it shows up as though the way length for A-B is JAI-BII-p, yet for B-An is IAI-BI1 + p.

Subsequently, the diffraction design is not, at this point centrosymmetric; sets of spots in the example that are connected by the focal point of symmetry, which are currently called Bijvoet sets, become inconsistent in force.

Given that the structure is known aside from the supreme configuration, one would then be able to figure the general powers of the Bijvoet sets for the R and S isomers, and by examination with analyze one can tell which will be which (Figure 5.7).

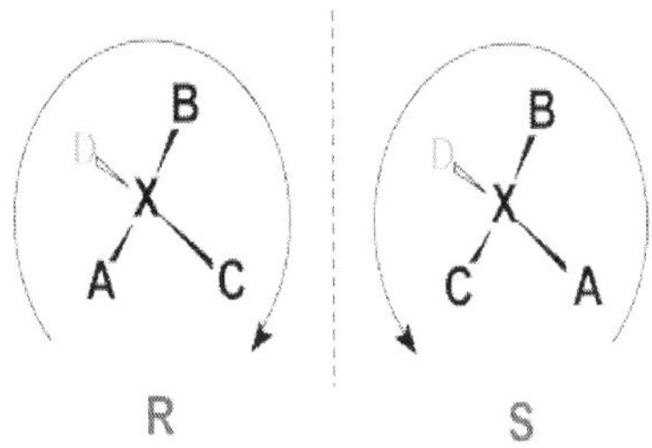

Figure 5.7. R S configuration.

Source: Image by Wikimedia.

Consequently, one can decide the feeling of chirality by utilizing X-beams on the retention edge of one of the atoms in the structure. For the most part, this is a moderately overwhelming molecule (sulfur or bigger) since the stage change increments with nuclear mass.

This standard was applied by Bijvoet et al. in 1951 (q.v.) by utilizing zirconium Ka X-beams in the X-beam crystallographic investigation of sodium rubidium tartrate. The (+)-tartrate anion was found to have the R, R configuration.

Since (+)-tartaric corrosive had been corresponded with numerous other chiral compounds, eminently the sugars, the assurance of its configuration denoted a significant achievement in stereochemistry. Before long a short time later the supreme configuration of the hydrobromide of the amino corrosive D-(–)-isoleucine was resolved using the stage slack at bromine presented by the utilization of uranium La radiation (Trommel and Bijvoet, 1954).

An intriguing variety of the Bijvoet technique has been utilized in the assurance of the total configuration of glycolic-d corrosive $CHDOHCO_2H$, got by enzymatic decrease of glyoxylic-d corrosive CDO-COH with NADH (nicotinamide adenine dinucleotide, diminished structure) within the sight of Liver alcohol dehydrogenase (Johnson et al., 1965).

The optically dynamic corrosive was changed over to its salt, which was then exposed to X-beam and neutron diffraction investigation. The X-beam investigation not out of the blue neglected to recognize hydrogen and deuterium (Seiler, Dunitz et al., 1984 for an effort to make the differentiation based on the distinction in the vibrational conduct of H and D). However, neutron diffraction discriminates between the two isotopes.

5.7. RELATIVE CONFIGURATION

Relative configuration can be used within the same molecule (that is intramolecularly, for example, cis-or trans-) or it can be used to intermolecularly (that is a pair of enantiomers which are having opposite configurations). For instance, consider 2-methylcyclohexanol.

There are two possible relative configurations that are based on the relative positions of the two substituents and whether they are on the same side or opposite faces of the cyclic structure.

Nevertheless, this system has two chirality centers and in this way, there are four configurational isomers with the absolute configurations shown. In simple words, this means that there are two possible configurations of the cis-structure.

The arrangement of atoms in an optically active molecule takes place according to the chemical interconversion from or to a known compound, is known to be a relative configuration. It is said to be relative, and the reason behind that, there is no way of knowing just having a look at a structure whether the assignment of (+) or (–) is correlated with respect to a particular enantiomer, R-or S-.

The reaction of an alcohol with TsCl (also known as 4-Toluenesulfonyl chloride) is known to take place with the retention of configuration, that is the group priority of the stereogenic center has not been changed or modified. The reaction of the tosylates with the nitrile takes place with the inversion. As an outcome, the group priority at the stereogenic center has been changed.

The expression "relative configuration" might be utilized in two types: relative configuration of a chiral focus of one compound or relative configuration of a few chiral focuses inside one compound.

Fischer projection equations for portrayal of compounds with more than one chiral focus. These equations have the downside, however, that they delineate the molecule in the dominated form.

Two different frameworks, the alleged Newman projection equations (Newman, 1955) and the "sawhorse" formulas show the molecules in their shocked form.

"Front visible" formulas fundamentally the same as the Newman projections have been utilized a lot before (Bischoff, 1891; Hermans, 1924). It likewise shows the interconversion of the different points of view.

It shows the supposed "flying wedge" and "crisscross" equations that are helpful to show the genuine compliance (just as configuration) of compounds having at least two chiral focuses (Newman and sawhorse formulas are generally helpful for compounds having close to two chiral focuses).

The flying wedge and crisscross equations have the upside of indicating the Molecular spine in the (genuine) amazed compliance and completely in the plane of the paper. The crisscross equation, although less complex, can be utilized distinctly for chiral focuses bearing a hydrogen atom, since this atom is expected.

The flying wedge formula doesn't experience the ill effects of this impediment and is helpful for the portrayal of any number of chiral focuses in a straight chain.

In a molecule having at least two indistinguishable chiral focuses, for example, tartaric corrosive, the way that two relating gatherings (here OH) are on a similar side of the atomic structure doesn't infer the nearness of a symmetry plane; this is as opposed to the circumstance in the Fischer projection.

The documentation for compounds having at least two chiral focuses should, on a fundamental level, be basic: Each inside is given its appropriate CIP descriptor alongside the standard positional number (locant).

L-threose is (2R, 3S)–2,3,4-trihydroxybutanal and D-glucose is (2R, 3S, 4R, 5R)–2,3,4,5,6-entahydroxyhexanal. This framework functions admirably when the compounds are settled (nonracemic) and their total configurations at each chiral focus (i.e., their outright just as relative configurations) are known.

One arrangement with racemates, or if just the family member yet not the outright configuration is known, the framework requires alteration. For racemates, the images RS or SR are utilized for the chiral focuses.

The most minimal numbered chiral focus is naturally given the image RS, which in this manner fills in as a source of perspective; let us call this chiral focus No. 1. The following chiral focus No. 2 is then signified by RS if, in the enantiomer where the No. 1 chiral focus is R, No. 2 is R additionally (and, similarly, in the enantiomer where No. 1 is S, No. 2 is S too).

Alternately, if the relative configuration is with the end goal that, when No. 1 is R, No. 2 is S (and, in this manner, assuming No. 1 is S, No. 2 is R), the configuration of the subsequent focus is indicated by SR.

The two racemic diastereomers are along these lines 1RS, 2RS and 1RS, 2SR, the previous meaning a racemic blend of 1R, 2R and 1S, 2S, the last of 1R, 2S and 1S, 2R. By method for models, let us consider the racemates of the open-chain types of threose and glucose.

Now, just the general configuration of a few chiral focuses, not the supreme configuration of any of them, is known. Since the RSISR framework just suggests relative configuration, it might be utilized right now.

Then again, a framework utilized by Chemical Abstracts (CA) might be utilized: The most reduced numbered chiral focus is discretionarily thought to be R and the configuration of the others is indicated as R* or S*, comparative with it.

So as to flag that the configurations are relative, not outright (i.e., the configuration of the first chiral focus is accepted discretionarily), all images are featured.

On account of the exceptionally visit event of compounds with two chiral focuses, scientific experts have needed to have words just as images for the general chirality of the two habitats required without independently determining the configuration of it is possible that one.

It is, obviously, conceivable to do this based on the CIP framework; therefore Bach and Prelog (1982) utilize the image l (for like) and u (for not at all like) when the CIP descriptors for the two chiral focuses resemble (R, R or S, S or RS, RS) or not at all like (R, S or S, R or RS, SR), separately.

The CIP framework is unambiguous in itself, it is not especially advantageous, on the grounds that it requires discovering the CIP descriptors for both chiral focuses before the l or u descriptor can be allotted.

What most scientific experts need is an "initially" assignment for an arrangement of two chiral focuses. It depends on the names of the four-carbon sugars erythrose and threose.

A_n and B (R # R'), isomers with the two indistinguishable ligands appended to a carbon chain on a similar side of the Fischer projection equation (as in erythrose) were called erythro-and those with the indistinguishable ligands on inverse sides (as in threose) were called threo.

5.7.1. Alpha and Beta Configuration

Alpha and beta are utilized in a specific manner for representing the overall configuration of starch molecules. They are utilized to depict the connection between the direction of one of the OH molecules on the cyclic compound versus the Fischer projection.

At the point when carbohydrates go from the straight structure (appeared in the Fischer projection) to the cyclic structure, one of the carbon iotas (C_5 in the direct structure and C_1 in the cyclic structure) can change the direction of the OH bunch during the progress.

If the direction remains on a similar side in the cyclic structure as in the linear structure, this is an alpha relationship. If the direction switches sides in the cyclic structure from the straight structure, this is a beta relationship.

5.8. CORRELATION BY STEREOSELECTIVE SYNTHESIS

A stereoselective combination includes presenting a stereogenic center in a molecule, so that one stereoisomer of the item is framed in inclination to the next or others. The blend might be either enantioselective or diastereoselective; in the previous case one of two enantiomers is framed specially, in the last case one of at least two diastereomers is shaped.

The two kinds will be considered here, however on account of diastereo selective blends it will limit to the situations where the recently presented stereogenic center is chiral. Cases including achiral stereogenic centers, for example, the decrease of 4-tert-butylcyclohexanone to cis-or trans-4-tert-butylcyclohexanol.

In an average case, response of an achiral forerunner with a chiral reagent (or within the sight of a chiral helper substance, called a chiral adjuvant) will give a chiral item.

In the event that the configuration of the reagent (or chiral adjuvant) is known, and if there is a one of a kind configurational relationship between the configuration of the reagent or adjuvant and that of the recently made chiral center, at that point the configuration of the last is corresponded with that of the previous.

The principle issue with this kind of connection is that the steric course of stereoselective "unbalanced" amalgamations is much more unsure than that of responses at chiral centers.

5.8.1. Prelog's Rule

This standard (Fiaud, 1977a) identifies with the course of lopsided union when a Grignard reagent is added to an a-ketoester (oftentimes a pyruvate) of a chiral auxiliary or tertiary alcohol, SMLC-OH (S, M, and L represent the little, medium-sized, and huge substituents, separately; S might be H).

Schematically, the response follows the stereochemical course (Prelog, 1953): If the ketoester is in the adaptation where the two carbonyl gatherings are antiperiplanar and in which the L bunch involves a similar plane as the two carbonyl gatherings and the alkyl oxygen, the Grignard reagent (RMgX) will approach the keto carbonyl capacity from the side of the S gathering of the two outstanding alcohol substituents.

This standard relates the configuration of the a-hydroxyester moiety (and accordingly of the a-hydroxyacid got by hydrolysis thereof) with that of the beginning alcohol SMLCOH fused in the a-ketoester.

Prelog's standard has been utilized to relegate the configuration of various alcohols optional ones (S = H) (Fiaud, 1977). The expansion contemplated is that of phenyl magnesium bromide to the pyruvate of the alcohol under investigation; this creates an atrolactate.

The atrolactic corrosive got after saponification is recognized polarimetrically as (S)-(+)-or (R)-$C_6H_5(CH_3)COHCO_2H$ and the configuration of the alcohol is concluded consequently (S acid+ R alcohol, and the other way around).

A few precautionary measures should be taken: It is significant that the saponification of the atrolactate be quantitative, in any case the outcomes might be distorted by motor goals.

Hence, if the transcendent ester item is that of the (+) corrosive, however the (diastereomeric) ester of the corrosive is saponified all the more quickly, it may show up, if saponification is deficient, that the corrosive prevailed in the item; a false end would consequently be drawn. This issue is kept away from by saponifying the ester totally.

5.8.2. Cram's Rule

Cram's rule refers to the diastereomer proportion shaped in the expansion of organometallic reagents to chiral ketones of the sort R'COCHXR, while Sharpless' standard arrangements with the proportion of enantiomers framed in epoxidation of RR'C=CR"CH_2OH with tert-butyl hydroperoxide and titanium tetra alkoxide within the sight of a R, R or S, S dialkyl tartrate.

Since the significant utilization of these principles is in stereoselective amalgamation. It may be noted, however, that the two standards (alongside Prelog's standard) have been utilized in the task of the outright configuration of $(CH_3)_3$ C-COH(CH_3)-CO_2H (Eliel and Lynch, 1987; Inch et al., 1968).

5.9. CONFIGURATION AND CHIRALITY

5.9.1. Molecular Geometry

In spite of the fact that stereochemical studies might be embraced (and in reality, were sought after for a long time) with no information on the genuine elements of molecules, increasingly significant outcomes are accomplished, if the geometrical parameters related with the molecules of intrigue are known.

Thus, these parameters will be characterized first. Molecules are made out of atoms, while atoms comprise of a core encompassed by various electron shells. The structures of the last lie in the area of quantum mechanics and need not be viewed as further here.

All in all, stereochemistry is worried about the general places of nuclear cores; the electronic structures related with the last are insignificant aside from the way that they do impact turnaround bonds.

In this manner turns around single bonds are quick in correlation with those about twofold bonds. At the point when a framework comprises of a solitary molecule, the geometry might be communicated regarding a solitary parameter, the van der Waals range (r).

Van der Waals* radii might be characterized as far as the vitality profile got when two non-reinforced nonpartisan atoms approach one another. As the internuclear separation diminishes, the underlying powerless appealing connection between the atoms is supplanted by a solid repugnance.

In a consolidated stage the internuclear separation (2r) relates to the base in the vitality separation bend, and consequently one method for deciding the van der Waals* sweep is through the estimation of the thickness of the dense stage.

As a first estimate, hence, atoms might be viewed as versatile balls with radii equivalent to the van der Waals' sweep r. The geometry of diatomic molecules might be depicted by the van der Waals' radii of the two fortified atoms (fA and fB) and their internuclear separation d, the supposed bond length, which is constantly shorter than the aggregate of the two radii. In triatomic molecules the atoms may lie on a straight line or structure a triangle.

Notwithstanding the three van der Waals* radii and the two bond lengths, a full portrayal of the framework likewise requires a meaning of the edge encased by the two bonds, the alleged bond point (Carbon dioxide, for instance, is a direct triatomic molecule, while water is non-straight.

In a four-atom molecule the atoms might be orchestrated in an orderly fashion (as in acetylene), in a typical plane (as in formaldehyde or boron trifluoride), or in three measurements (as in smelling salts).

The interconnections between the atoms may be unique, straight or spread. In direct four-atom molecules, or in four-atom segments A–B–C–D of a chain, another parameter, the torsion or dihedral edge q > must be presented.

While the groups of three A–B–C and B–C–T > might be portrayed by the two bond points, 6± and 02, the connection between these sets of three is given by the torsion edge cp. Torsion edges might be promptly envisioned through Newman projections; in such a projection a fanciful plate is set over the bond B–C, the molecule being then seen along the course of this bond.

There lies a portion of the torsion edges that must be supplanted by bond points. The geometrical parameters of molecules are firmly identified with the vitality substance of the last mentioned.

Hence, in considering molecules as flexible chunks of van der Waals' range r, we are recommending that the electron shells related with these atoms are generous and fit for infiltration just by the consumption of a lot of vitality.

Thus, interatomic bond lengths have alleged normal or balance esteems (e.g., 154 pm on account of C–C single bonds) when the vitality of the bond is at least. The extending or shortening of bonds comparative with this balance esteem includes an expansion in vitality.

Additionally, three-molecule clusters have minimum energy balance bond points (e.g., 109–111° for C–C–C single bonds) and direct chains of four molecules have ideal qualities for the relating dihedral edges.

As a matter of fact, it might be expressed that, except for torsion edges, the geometrical parameters of molecules are very much characterized, for example, deviations from balance esteems are little. It is implicitly accepted in organic science that indistinguishable groupings of atoms have an indistinguishable geometry.

It is underestimated that in propane the C_1–C_2 and C_2–C_3 bond lengths are equivalent. This is equivalent to recommending that atomic geometry is constrained by symmetry, as such Molecular symmetry is a significant factor in deciding the genuine properties of molecules.

5.9.2. Molecular Symmetry

The symmetry of any object might be depicted as far as its components of symmetry. Two basic components of symmetry exist, the plane and the axis of symmetry, and what's more it is valuable to think about a composite component, the revolution reflection axis (Figure 5.8).

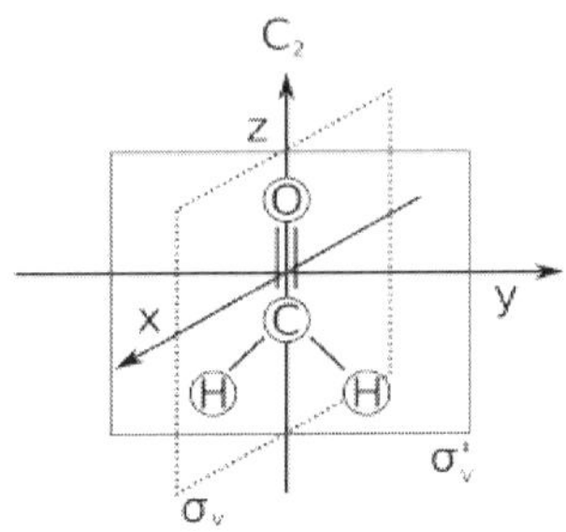

Figure 5.8. Molecular symmetry.

Source: Image by Wikipedia.

Using planes of symmetry (or mirror planes) (signified by the image an) objects can be cut up into parts which are identified with one another as perfect representations. Note that it is not important for the two parts to be superimposable.

Instances of items having a plane of symmetry as the main symmetry component are coffee cups, while deuterium hydroxide might be taken for instance of a regular molecule right now.

5.10. CONCLUSION

The symmetry and configuration are the two essential aspects of the organic compounds. On the basis of these two factors the compounds and their properties can be understood. Like the symmetry of a chemical compound helps in describing the chemical property of a given molecule or compound.

The chemical structures formation is dependent on the configurational aspect that has been described here. The symmetry also helps in the classification of molecules as on the basis of the symmetry of the chemical compound the molecules can be classified, and their groups can be decided.

The relative and the absolute configurations have been properly evaluated here in order to determine the importance of configuration for chemical compound formation. In organic chemistry, stereochemistry plays an essential role, the different aspects of stereochemistry have been evaluated here.

Symmetry of the chemical compounds helps in the study of chemical structure of the chemical compounds. Their properties are studies by understanding the symmetrical behavior of the chemical compounds.

The characteristics of the symmetry of the chemical compounds includes factors for evaluation such as identity symmetry, plane of symmetry, inversion symmetry and its center, improper rotation symmetry, and n-fold improper rotation.

The symmetry of the chemical compounds, as discussed in the chapter, is very essential for the classification of the compounds. The compounds are grouped together based on their symmetrical properties.

Every compound that falls in the same category has few properties in common, it allows them to be used for specific scientific and testing purpose.

Similarly, configuration plays an equally essential role in the classification and evaluation of chemical compounds. The properties such as molecular point groups are described by analyzing the configuration of chemical compounds.

In relative configuration the Newman projection equations and the sawhorse formulas have been considered as a framework for determining the state of the molecule.

Also, the Bijvoet Method is evaluated that describes the process of determining the absolute configuration. Additionally, Prelog, and Cram's rules have been described so that stereoselective synthesis can be understood. All the major aspects related to the configuration and symmetry aspects of the organic compounds have been described in an efficient manner.

REFERENCES

1. Chemistry LibreTexts, (2019). Cram's Rule and Prelog's Rule. [online] Available at: https://chem.libretexts.org/Bookshelves/Organic_Chemistry/Book%3A_Logic_of_Organic_Synthesis_(Rao)/02._Rules_and_Guidelines_Governing_Organic_Synthesis/Cram's_Rule_and_Prelog's_Rule (accessed on 18 August 2020).
2. Eliel, E. L., & Wilen, S. H., (2008). Stereochemistry of Organic Compounds. John Wiley & Sons.
3. Nógrádi, M., (2013). Stereochemistry: Basic Concepts and Applications. Elsevier.
4. Reusch, W., (n.d.). Symmetry. [online] Www2.chemistry.msu.edu. Available at: https://www2.chemistry.msu.edu/faculty/reusch/VirtTxtJml/symmetry/symmtry.htm (accessed on 18 August 2020).
5. Study.com. (n.d.). Relative Configuration in Organic Chemistry: Definition and Examples. [online] Available at: https://study.com/academy/lesson/relative-configuration-in-organic-chemistry-definition-examples.html (accessed on 18 August 2020).

Chapter 6

Separation of Stereoisomers, Resolution, and Racemization

CONTENTS

In the chapter separation of stereoisomer's, resolution, and racemization, the various process, strategies, and methods in separation is discussed. The chapter also explain the separation of enantiomers by crystallization. Some of the popular methods in this such as crystal picking triage, conglomerates, and preferential crystallization is discussed.

The chapter also discuss the chemical separation of enantiomers through diastereomers. The chapter also through some light nonracemic mixtures and the resolution of enantiomers. In addition, the chapter also highlight the resolution of racemic mixtures of amino acids in which natural α-amino acids, the isoelectric point and other natural amino acids are discussed. In the end, the chapter explain in detail the concept of racemization.

6.1. INTRODUCTION

The creation of non-racemic samples of chiral organic compounds either from racemic precursors or from achiral by adopting any viable means is known as optical activation.

It is usually seen that optical activation is not considered as satisfactory expression by many. It clearly positions as the formation of a physical property (which is called optical activity), while on the other hand, the portrayal of the process ought appropriately to highlight the enantiomerically enhanced composition of the preferred product quite apart by using rotation as a probe.

Moreover, this term has lot of shortcomings primarily because of the fact that large number of samples of enantiomerically enhanced with chiral substances may carry meagre or even in some cases zero rotation. Unfortunately, it results in nonexistence of any alternative expression.

If the isolation of chiral substances from either animal sources or plant sources and their alterations is debarred, there are majorly two key sets of optical activation methods: stereoselective syntheses and resolutions.

A resolution refers to the parting of a racemate into its two enantiomer constituents. In methods of resolution, the departure point is known by the name racemate; thus, the highest amount of yield of individual enantiomer is approximately 50 %

There are various methods that may involve in resolution methods (a) physical processes only or (b) chemical reactions. In resolutions that are governed by the chemical reactions, it is generally seen that there is intervention of either diastereomeric products or diastereomeric transition states.

While using such types of methods, one can be benefitted from the operations of either kinetic control or thermodynamic. It is widely assumed that whatever be the nature of the chemical alteration that occurs in context with a resolution, the reaction is either rescindable or, in some other case, the process eventually triggered back to the initial material parted into its enantiomer components.

Hybrid processes which includes the resolved products are also considered as the derivatives of the starting racemates and, in this sometimes only one of the enantiomers is retrieved from the parting. All these processes do not come into the above discussed category but however known by the name resolutions. In this chapter various resolution methods are being discussed.

6.2. SEPARATION OF ENANTIOMERS BY CRYSTALLIZATION

6.2.1. Crystal Picking Triage

When racemate's crystallization resulted in creation of a conglomerate then, by meaning, the element is said to be impulsively resolved. This means that during the process of racemate crystallization under equilibrium conditions, whether this occur immediately at a sluggish rate or at a rapid rate when persuaded with seed crystals, both enantiomers of the substance accumulated in identical amount as enantiomorphs crystals.

The physical sorting of conglomerate crystals into segments whose outcomes are levorotatory and dextrorotatory (as computed by circular dichroism, polarimetry, or CD) is known as triage. It is usually seen that palpation of the crystal handedness can ramp up the development and this is achievable when there is well formation of crystals and this is endowed with hemihedral faces.

Two such examples of resolutions as that stated by Pasteur (1848; Kauffman and Myers, 1975) and that of asparagine (Piutti, 1886). Various stratagems for differentiating crystals tended to arise the need of oppositely signed solutions without depending on the existence of hemihedral faces and this have been discussed by Collet, Jacques, and Wilen (1981) and Kaneko et al. (1974).

When there is availability of a crystal of a size 2 mm long, a small piece can be separated with the help of a razor blade. Them the handedness of the crystal is figured either chromatographically orthotopically after the dissolution of small piece in a suitable solvent. Solutions of distinctive crystals of the similar category of chirality may then be united for retrieval of pieces of the crystal and for grouping with the consequent primary crystals.

This task of operation is usually lengthy or tiresome in nature. When there is proper information about the chirality sense is accessible from crystal shapes, then resolution with the help of triage becomes more feasible.

While such kind of widely recognized illustration is rac-hydrobenzoin (Fieser, 1964), examples of resolution with the help of triage consistently gathered in the literature, even if only to attain just adequate nonracemic sample for study of configuration (Asaoka et al., 1988; Allen, Green, et al., 1983; Ireland, Johnson, et al., 1970; Dossena et al., 1983; Overberger and Labianca, 1970).

One should not underestimate the potential of resolving trivial amounts of a racemate in this very straight forward and efficient way that deliberately avoid the use of chiral auxiliary reagents. Neither the use of a hand lens norhemihedral facets is necessary for the accomplishment of undergoing the resolutions by straight forward crystallization of enantiomer mixtures.

6.2.2. Conglomerates

It is widely recognized that in a way for resolutions of enantiomer mixtures to be feasible without the use of enantio selective chromatographic columns or of chiral reagents, the racemate should consist of different or distinct number of item (refereed as conglomerates) under the conditions of the crystallization. While only limited number of conglomerates have been explained so far (Jacques et al., 1981a, p. 58), it is more likely that there is presence of many more in the domain of chiral organic compounds.

In a sample of 1308 neutral compounds drawn from the Beilstein Handbook, it was revealed that conglomerates exist between 5 to 10% of the totality of chiral organic solids (Leclercq, Jacques, and Brienne, 1981b).

Even if a given racemate does not form or cause to form crystals as a conglomerate, then there are chances to convert it reversibly into a derivative. For instance, some of the commonly used amino acids are resolved impulsively, many of the known derivatives are conglomerates.

Leucine, alanine, and tryptophan cause to form crystals as racemic compounds, however their benzene sulfonate salts are conglomerates. In addition, it is also important to note that histidine hydrochloride is a conglomerate, and there are some other instances (Jacques, Collet, and Brienne 1980). Overall, it was noted that the rate at which such derivatization conducive to form conglomerate is drastically amplified when the derivative is a salt instead of a covalent compound (Leclercq, Jacques, and Brienne, 1981).

Thus, while (+)-tartaric acid form into crystals as a racemic compound, the initial substance ever determined, Pasteur's salt (tartrate.4H_2O, racemic NaNH), is a conglomerate which is below the degree of 28°C as is "Seignette's salt" (tartrate.4H_2O, racemic NaK) below –6°C; these double salts are determined impulsively.

These two discussed instances also demonstrate the likelihood of converting certain conglomerates into fewer highly solvated racemic compounds because of high level of temperature (Jacques et al., 1981a, pp. 131, 201) and vice versa.

Two more illustrations will determine the utility in resolutions of derivatives that unveil conglomerate behavior. The alcohol 2 and the mine 1 are both liquids at temperature of room level. The established determinations of these substances through crystalline diastereomeric derivatives is not mainly problematic, although a particular quantity of tedium may be linked to the wanted recrystallizations. However, it is generally seen that solid derivatives 3–5 are conglomerates.

The implications of this information are as follows: Crystallization of each of the racemates (3–5) under the balanced conditions (gradually by evaporation or with cooling even in the existence of racemic seeds) augment the need of a globally racemic precipitate from which it can be feasible to recover (or dispersed) wisely well-formed homochiral crystals of each chirality sense.

Such crystallization parallels to a classic Pasteurian resolution in its original form. Seeding with homochiral crystals of one of the enantiomers shift the procedure of non-equilibrium crystallizations. Here, a greater quantity of one of the enantiomers of the substance may be placed in a mother liquor whose optical rotation may be altered during the process of crystallization.

Favored crystallization is a practical form or version of such nonequilibrium crystallizations that take place under fully controlled conditions intended at averting the impulsive crystallization of the unwanted enantiomer during the organized crystallization of its chiral partner (Figure 6.1).

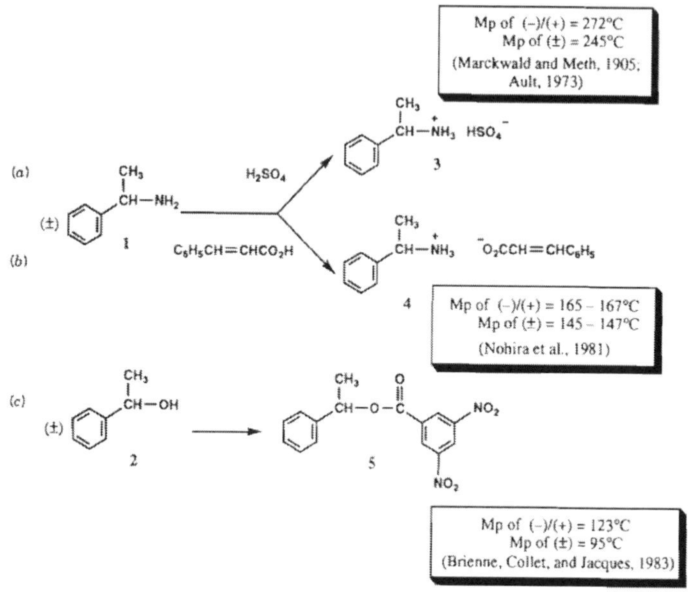

Figure 6.1. Derivatives exhibiting conglomerate behavior.

Source: Image by Wiley inter-science publication.

6.2.3. Preferential Crystallization

When resolution of a racemate is originated by inoculation of a supersaturated or saturated solution of the racemate with seeds (such as crystals) of one of the enantiomers, crystallization occur in a non-equilibrium process known as special crystallization, as it is the only existing seeded enantiomer that deposits.

It is often seen that the experiment of crystallization occurs on aqueous solutions of sodium chlorate ($NaClO_3$), which is an achiral substance forming crystals in the chiral cubic space group P2, P3, have intensely determined the differences between crystallization under nonequilibrium and equilibrium conditions.

Undisturbed crystallization (also refer as equilibrium conditions) accentuate the need of identical numbers of enantiomeric crystals, that is, to a racemic, or closely racemic, solid. While, on the other side, if process of forming crystals is carried out with constant stirring (known as kinetic control) without the need of seeding, the solid product is almost enantiopure (>99%) with an allocation of 14 predominantly (–) and18 predominantly (+) crystal batches found in 32 crystallization trials (Kondepudi et al., 1990; see also McBride and Carter, 1991).

It is important to note that the preferential crystallization works only for those substances that are conglomerates (Inagaki, 1977). More specifically, it could perform better with compounds present either as racemic compounds oars conglomerates within a desired temperature level, given that the racemic compound does not transform into crystals during the operation.

The latter qualification is to make sure the likely presence of metastable racemic compounds that may hinder a preferential crystallization (Collet, Brienne, and Jacques, 1980). Thus, it is generally seen that the histidine HC1 is resolvable by preferential crystallization but only when the temperature level is consistently above 45°C.

6.3. CHEMICAL SEPARATION OF ENANTIOMERS VIA DIASTEREOMERS

6.3.1. Formation and Separation of Diastereomers, Resolving Agents

The huge amount of recorded resolutions has significantly been impacted by transformation of a racemate into a mixture of diastereomers. In such kind of reaction, the substrate to be resolved need to be fixed with one enantiomer of a chiral substance (which is the resolving agent). The first such resolution, demonstrated by Pasteur in 1853, is discussed in Figure 6.2 (Jacques et al., 1981a).

Diastereomer pairs being arranged in context resolutions may be covalent, ionic (diastereomeric salts), inclusion compounds or charge transmission complexes. The large number of resolutions resolved by diastereomers (particularly with diastereomeric salt mixtures) have been rely upon solubility variances of solids; although, in the contemporary literature, separation of covalent diastereomer is dependent on chromatography in all of its variations are used with pronounced frequency (Figure 6.2).

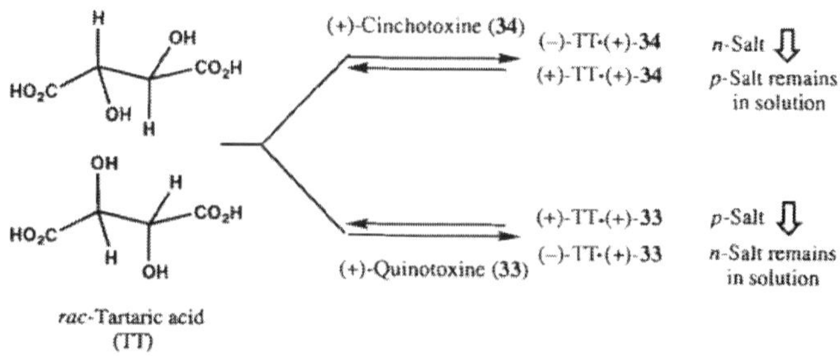

Figure 6.2. First resolution via diastereomers.

Source: Image by Wiley inter-science publication.

It is generally seen that chromatography has freed resolutions from the restriction of dependency on crystallization as the method on which diastereomer separation has been largely depended. Because of this, resolutions in general have shown more success rate at present compared with the past results. Not rarely oily covalent diastereomer mixtures ultimately form into crystals and their resolution then be function in a more customary way by getting the benefits of solubility differences.

It is now possible to engage in the separation of diastereomeric salt mixtures. Because of this subsequent impact between covalent and ionic structure and the variety of approaches to separate diastereomer mixtures, it has been decided to not to handle resolving agents individually irrespective of whether they form ionic or covalent diastereomer mixtures.

The variance in steadiness of diastereomers is another basis for their separation. It is usually seen that transient diastereomeric species are created in chromatographic columns as there is interaction of flowing racemate samples with achiral stationary phases or with chiral stationary phases in the existence of nonracemic chiral mobile phases. Some of the notable characteristics of a good resolving agent are discussed as follows (Wilen, 1971):

a. Stability of supply;
b. Ready availability;
c. Consistency in use and in storage;
d. Ease of preparation and economical;
e. Low molecular weight;

f. Ease of reuse and recovery;
g. Accessibility of both enantiomers;
h. Availability in high enantiomeric purity;
i. Adequate solubility; and
j. Low toxicity.

6.3.2. Separation via Complexes and Inclusion Compounds

Compounds that lack conventional functionality (such as chiral arenes and alkenes, phosphines, and sulfoxides), for which there are limited alternative resolution routes and these may be resolved by inclusion into diastereomeric metal complexes or through any sort of reaction with chiral T-acids (Jacques et al., 1981a).

According to Cope et al., it was recognized that the resolution of trans-cyclooctene by integration of the alkene enantiomers into diastereomeric trans platinum (0) square planar complexes (Cope et al., 1962, 1963) are detachable through low level temperature of crystallization. Other dienes, alkenes for instance, spiro-hepta–1,5-diene (McKervey, Hulshof, and Wynberg, 1974), arsines, allenes, sulfoxides, and phosphines have been resolved analogously (Figure 6.3).

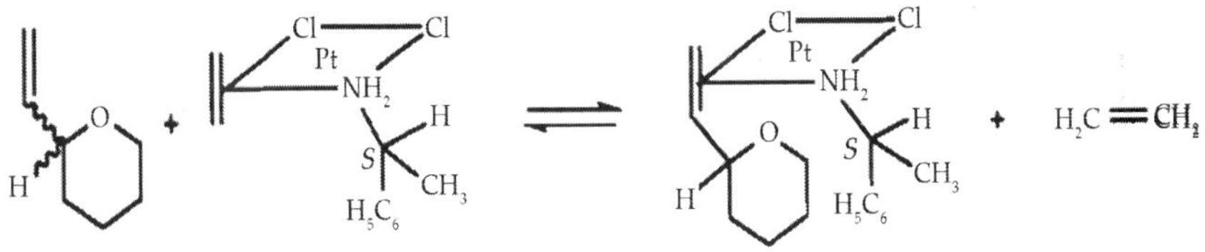

104

Figure 6.3. Resolution of chiral alkenes via neutral platinum coordination compounds.

Source: Image by Wiley Interscience Publication.

Inclusion of molecules within others is a commonly adopted phenomenon and the applicability of the "inclusion method" to resolutions, as demonstrated by Schlenk (1965), is no longer considered more viable. Two prominent classes of inclusion compounds:

- The cavitates (Cram, 1983) in which it is possible to resolve the chiral substrate (the guest molecule) either partially or completely enclosed within a second chiral substance (the host molecule and resolving agent) that is endowed with a chiral cavity

- Those in which guest molecules are bounded by numerous molecules of the resolving agent forming a channel or a cage; these are usually referred as clathrates (Weber, 1987); Arad-Yellin, Worsch, and Vogtle (1987); Green, et al. (1984); Tsoucaris (1987); and Toda (1987).

It was observed that the preferential inclusion of one substrate enantiomer is because of the van der Waals and hydrogen bonding attraction between host and guest molecules. Although, in addition, there should be appropriate balance between the shape and size of the guest molecule and a resultant shape and size of the host cavity. While there could be variation in the shape and size of guest molecules if they are conformationally flexible in nature.

There may be presence of more play in the space in the host molecules especially when lattice inclusion compounds are concerned. Parting of the diastereomeric inclusion compounds may vary on number of factors such as differential solubility or stability. Recovery of the resolution substrate may engage series of activities such as dissolution, melting, or extraction of the enantiomerically enriched guest.

6.3.3. Chromatographic Resolution

The large number of different categories of chromatographic resolutions reported in the literature have been analytical in nature, which means their primary goal is to determine stereoisomer and, in specific, enantiomer composition.

This section is basically concerned with preparative chromatographic resolutions. Preparative chromatographic resolutions are in most of the instances carried on covalent diastereomer mixtures. In majority of the cases, chromatographic resolution of diastereomers on silica gel able to achieve its objective even with the simple and prompt flash chromatography technique (Still et al., 1978) with the supervision of initial thin-layer chromatographic (TLC) experiments in which the only variable is solvent composition and others are fixed (e.g., Comber, and Brouillette, 1987).

The resolution of gossypol as the bases of Schiff with numerous amino acid esters (for instance methyl L-phenylalaninate) by TLC on silica gel could be produced up to distinct 5 g (and even bigger in some cases) samples on silica gel columns (Tyson, 1988).

The chromatographic resolution of alcohols as diastereomeric carbamates (usually made with isocyanate) in a computerized liquid chromatograph (on approximately 1-gram samples) has been defined (Hoekstra and Pirkle, 1974).

The simplicity with which covalent diastereomers are parted chromatographically highlight the importance of the anisometric character of such isomers. In case, there is breakdown of such chromatographic resolutions, then it is most likely because of not forming into the structures an enough number of intermolecular binding sites (or sites allowing adequately strong intermolecular binding) to authorize discriminately to portray itself as the diastereomeric molecules voyage along the stationary phase.

The reader is reminded of an important limitation of such resolutions: separation via covalent diastereomers will necessarily be incomplete if the resolving agent is not enantiomerically pure (see p. 241). When carried out on chromatographic columns of high efficiency (>10,000 theoretical plates), baseline separations are easily achieved when separation factors attain or exceed a = 1.2 [a = (t2–tl)l(tl–to) where t, and t, are the retention times of the second and first diastereomers eluted, respectively, and to is that of an unretained substance, e.g., pentane].

The choice of resolving agent was primarily based on the existence of a suitably placed hydroxyl group that expedites the hydrolysis of the diastereomeric amides as a result of their parting, probably by neighboring group participation.

However, in theory, preformed ionic diastereomer mixtures can be parted with the help of chromatography (Knox and Jurand, 1982), in practice ion-pair chromatographic resolutions have embrace the supplement of nonracemic counter ions to the mobile phase. (+)-10-Camphorsulfonic acid has been worked as the mobile phase additive in the resolution of (–)-quinine and chiral amino alcohols.

HCl has been used as additive in the resolution of (*)-10-camphorsulfonic acid. These separations in most of the cases is analytical in scale (Pettersson and Schill, 1986, 1988; Szepesi, 1989). The most powerful and flexible chromatographic resolutions are those that can be occur on enantioselective stationary phases.

6.4. RACEMIC MIXTURES AND THE RESOLUTION OF ENANTIOMERS

It is well known that chiral compounds that are basically synthesized from a chiral starting materials and reagents are generally racemic in nature (for instance a 50:50 mixture of enantiomers). A process in which the race mates separated into their component enantiomers are known as resolution.

Since enantiomers carries their own unique physical properties, such as melting point and solubility, resolution is a very challenging task. While on the other hand, Diastereomers, have diverse physical properties, and this exclusive characteristic of diastereomers allow them to achieve resolution of racemates.

When a racemate is reacted with an enantiomerically pure chiral reagent, it gives a mixture of diastereomers that can be easily separated.

For instance, if a racemic mixture of a chiral alcohol comes in reaction with a enantiomerically pure carboxylic acid, then it resulted in a mixture of diastereomers: in this case, due to the fact that the pure (R) enantiomer of the acid was come into use, the product is a mixture of (R-S) and (R-R) diastereomeric esters, which can, in practice, be parted by their diverse physical properties.

Subsequent hydrolysis of each estranged ester will produce the 'resolved' (enantiomerically pure) alcohols. The technique that come into use in this is known by the name 'Moscher's esters,' after Harry Stone Moscher, a chemist and researcher at Stanford University pioneered the method (Figure 6.4).

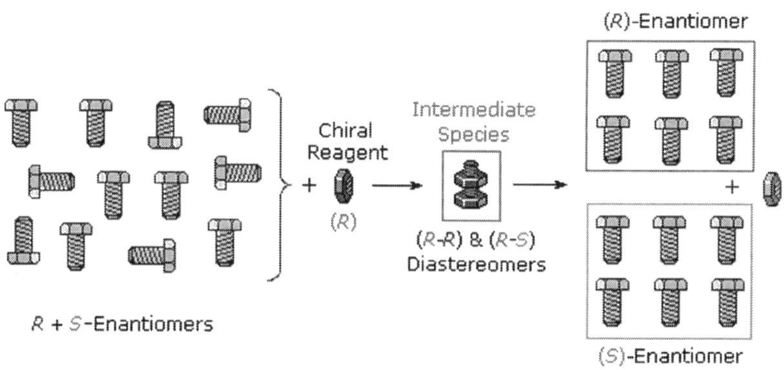

Figure 6.4. Moscher's esters technique.

Source: Image by Libretexts.

There are variety of physical and chemical reactions available, including salt formation that can be used in achieving the diastereomeric intermediates essential for separation. Figure 6.4. depicts this common principle by displaying how a nut equipped with a right-handed thread (R) could work as a "reagent" to separate and differentiate a mixture of left-handed as well as right-handed bolts of similar weight and size. Only the two right-handed partners can interact to give a fully-threaded intermediate, so separation is fairly simple.

The resolving moiety, i.e., the nut, is then removed, leaving the bolts separated into their right and left-handed forms. Chemical reactions of enantiomers are normally not so dramatically different, but a practical distinction is nevertheless possible.

Because the physical properties of enantiomers are identical, they seldom can be separated by simple physical methods, such as fractional crystallization or distillation. It is only under the influence of another chiral substance that enantiomers behave differently, and almost all methods of resolution of enantiomers are based upon this fact. These are included in a discussion of the primary methods of resolution.

6.5. THE RESOLUTION OF RACEMIC MIXTURES OF AMINO ACIDS

Proteins, derived from the Greek word *proteios*, meaning first, are a category of organic compounds that present in each and every living cell, which play very important role for its overall functioning. In the form of skin, callus, hair, muscles, cartilage, ligaments, and tendons, proteins hold together, provide, and protect, structure to the body of a multi-celled organism. In the form of hormones, enzymes, globulins, and antibodies, they govern, monitor, protect, and catalyze the body chemistry.

In the form of myoglobin, hemoglobin, and various lipoproteins, they are responsible for transporting the oxygen and other critical substances within an organism.

Proteins are always considered beneficial, and play an essential role in diet of all animals. Humans can fall ill if they do not take adequate amount of protein in their day to day life, the disease commonly known as kwashiorkor is an extreme form of protein deficiency. Protein based vaccines and antibiotics help in evacuating such kind of problems and human beings in general protect and warm their bodies with clothing and shoes that are usually of protein in nature (for instance; silk, wool, and leather).

The deadly properties of protein venoms and toxins is less widely appreciated. Botulinum toxin A, which is mainly extracted from Clostridium botulinum, is conceived as the strongest poison known till now. Based on the past and recent toxicology studies, a meagre amount of this toxin would be as hazardous as it can kill 20% of the entire globe population.

The toxins produced by diphtheria and tetanus microorganisms are closely as poisonous. A wide category of extremely toxic proteins or peptides would also comprise the venoms of many ricin and snakes, the toxic protein found in castor beans.

Despite the extensive diversity of their physiological function and variation in physical properties--silk is an adaptable fiber, horn a tough rigid solid, and the enzyme pepsin water soluble crystals--proteins are mostly identical in molecular structure to warrant treating them as a single chemical family. In comparison with lipids and carbohydrates, the proteins are comparatively distinctive in fundamental composition.

The lipids are in majority of the cases hydrocarbon in nature, generally being 75 to 85% carbon. Carbohydrates are approximately 50% oxygen, and like the lipids, usually have less than 5% nitrogen (often none at all).

While on the other hand, both proteins and peptides are composed of 15% to 25% nitrogen and about a similar quantity of oxygen. The difference between proteins and peptides can be easily observed by looking at their size. Peptides are in a sense small size of proteins, having molecular weights which is not beyond 10,000.

6.5.1. Natural α-Amino Acids

Hydrolysis of proteins derived by boiling aqueous acid or base produce a collection of small molecules recognized as α-amino carboxylic acids. There are more than 20 such components which have been isolated so far, and the most typical one is listed in Table 6.1.

Table 6.1. Physical Properties of Selected Acids and Amines

Compound	Formula
Isobutyric Acid	$(CH_3)_2CHCO_2H$

Lactic Acid	$CH_3CH(OH)CO_2H$
3-Amino-2-Butanol	$CH_3CH(NH_2)$ $CH(OH)CH_3$
Alanine	$CH_3CH(NH_2)CO_2H$

Source: Table by Libretexts.

Those amino acids which are of green colored names are crucial diet components, since they are not synthesized by human metabolic processes. The best food source of these nutrients is protein, but it is important to recognize that not all proteins have equal nutritional value.

For instance, peanuts have high amount of protein as compared to eggs or fish, but the percentage of vital amino acids in peanut protein is just only a 33% of that from the two other sources. For all these reasons, this will serve as a greater basis discussing the structures of proteins and peptides, each amino acid is allotted a one- or three-letter abbreviation.

There is need to note some of the essential characteristics of these amino acids. With the exception of glycine, all of them are chiral, and with the exception of proline, most are 1°-amines. It is generally seen that the configurations of the chiral amino acids are almost identical when written as a Fischer projection formula, and this was expressed as the L-configuration by Fischer.

The R-substituent can be defined as the left structural element that differs from one amino acid to another, and in proline R is a three-carbon chain which linked the nitrogen to the alpha-carbon in a five-membered ring.

When the notation of Cahn-Ingold-Prelog (CIP) applied, majority of the natural chiral amino acids, apart from cysteine, have an S-configuration. It is important to note that for the first seven compounds in the left side of column the R-substituent is a hydrocarbon, while the preceding three entities in the left column side have hydroxyl functional groups.

The first two amino acids in the right column side incorporate sulfide and thiol groups correspondingly. Arginine and Lysine have basic amine functions in their side-chains; tryptophan and histidine have less simple nitrogen heterocyclic rings as substituent's. Ultimately, carboxylic acid side chains are substituents on glutamic and aspartic acid, and the latter two compounds in the right side of column are their corresponding amides.

The amino acids formula transcribed above are ingenuous covalent bond representations relying on the prior knowledge of mono-functional analogs. However, the formulas are inaccurate. This can be validated by engaging in comparison of the physical properties listed in the following table. All compounds discussed in the table are of similar size, and all have adequate to exceptional water solubility. There are three compounds out of which first two are simple carboxylic acids, while the third one is an amino alcohol.

All of the compounds have the potential to get soluble in organic solvents (for instance; ether) and have comparatively low melting points. It is further important to note that the carboxylic acids have pK_a's near 4.5, and the related acid of the amine has a pK_a of 10.

In addition, the simple amino acid alanine is the last entry. By contrast, it carries extreme melting characteristics (with decomposition), which is insoluble in organic solvents, and a squillion times weaker as an acid when compared with common carboxylic acids.

All these points signal towards internal salt creation by a transfer of proton from the acidic carboxyl function to the basic amino group. The resultant structure of ammonium carboxylate structure, widely known as a zwitterion, is also reinforced by the characteristics of spectroscopic of alanine.

As it can understand from its ionic character, the alanine zwitterion is of high melting in nature, and also insoluble in non-polar solvents and equipped with the acid strength of a 1°-ammonium ion. To the right side is a J-mol display of an L-amino acid. The model can be changed to its zwitterionic form by pressing the right button underneath the display.

There are various examples of some definite amino acids that can be considered in their preferred neutral zwitterionic form. It is important to note that in lysine the amine function extreme from the carboxyl group and is more basic compared with the alpha-amine. Subsequently, the ammonium moiety which is positively charged is mainly created at the chain terminus and attracts towards the negative carboxylate, stemming in a coiled conformation.

Since amino acids, as well as proteins and peptides, incorporate both basic as well as acidic functional groups, the predominant molecular species which exists in an aqueous solution will primarily rely on the pH of the solution. In a way to get an estimation of the ionic and molecular species that exists in aqueous solutions at different pH's level, Henderson-Hasselbalch Equation play very important role and is discussed as follows.

In the equation discussed below: the pK_a defines the acidic level of a specific conjugate acid function (HA). When the pH of the solution is equivalent to the pK_a, the concentrations of $A^{(-)}$ and HA must be equal (log 1 = 0).

$$\text{Henderson – Hasselbalch Equation}: pK_a = pH + \frac{[HA]}{[A^-]}$$

The titration curve for alanine, as discussed below, displays this relationship. At a pH lower than 2, both the amine and carboxylate functions are protonated, which resulted in developing a net positive charge by alanine molecule. At a pH beyond 10, the amine presents as a neutral base while the carboxyl as its conjugate base, resulted in causing an alanine molecule to develop a net negative charge.

At intermediate pH's the proportion of zwitterion increases, and at a characteristic pH, referred as isoelectric point (pI), the positively and negatively charged molecular species are existing in equal proportion. This behavior is general for simple (difunctional) amino acids.

At the early fully protonated state, the pK_a's of the range of acidic functions varies from 8.8 to 9.7 for-$NH_3^{(+)}$ and 1.8 to 2.4 for-CO_2H. in addition, the range of isoelectric points varies from 5.5 to 6.2. Titration curves demonstrates the neutralization of these acids by the variation in pH and added base during the titration.

The allocation of charged species in a sample can be presented experimentally by witnessing the activity of solute molecules in an electric field, with the help of electrophoresis technique. For these types of experiments, an ionic buffer solution is integrated into a solid matrix layer, composed of a cross linked gelatine or of paper like substance.

A meagre proportion of the amino acid, protein or peptide sample is positioned near the middle of the matrix strip and an electric potential is placed at both side of the strip, as describes in the diagram below.

The solid structure of the matrix delays the progress of the diffusion of the solute molecules, which remain in existence where they are inserted, unless acted upon by the electrostatic potential. In the below described example, there are four different amino acids assessed concurrently in a pH 6.00 buffered medium.

At pH 6.00 level, isoleucine, and alanine presence on average as neutral zwitterionic molecules, and are not majorly impacted by the electric field. Arginine refers to a basic amino acid. Both base functions present as "onium" conjugate acids in the pH 6.00 matrix.

Therefore, it is noted that solute molecules of arginine take with them high amount of positive charge, and they move toward the cathode. The two carboxyl functions in aspartic acid are both ionized at pH 6.00, and solute molecules which are the negatively charged move toward the anode in the electric field. Structures for these species can be observed to the right of the display.

It is important to note that the outcome of this experiment primarily relies on the pH of the matrix buffer. In case, there is need to reiterate the electrophoresis of these compounds at a pH level of 3.80, then it is more likely that the aspartic acid would remain at its point of origin, while on the other hand, the amino acids would be shifted toward the cathode.

By overlooking the variations in molecular size and shape, the arginine would transfer at a rapid rate, twice as fast as the isoleucine and alanine because its solute molecules on usual holds a double positive charge.

As it is earlier observed, the titration curves of manageable amino acids exhibit two inflection points, one because of the highly acidic carboxyl group (pK_a^1 = 1.8 to 2.4), while the other for the fewer acidic ammonium function (pK_a^2 = 8.8 to 9.7). It is generally seen that for the 2°-amino acid proline, pK_a^2 is 10.6, predicting the greater basicity of 2°-amines.

There are some of the amino acids that possess extra basic or acidic functions in their side chains. A comprehensive list of these compounds is listed in Table 6.2 presented: A third pK_a, signifying the basicity or acidity of the extra function, is described in the fourth column of the table. The pI's of these amino acids (discussed in the last column) are usually very significantly compared with those noted above for the simpler members.

Table 6.2. pKa Values of Polyfunctional Amino Acids

Amino acid	α-CO_2H pK_a^1	α-NH^3 pK_a^2	Side Chain pK_a^3	pI
Arginine	2.1	9.0	12.5	10.8
Aspartic acid	2.1	9.8	3.9	3.0
Cysteine	1.7	10.4	8.3	5.0
Glutamic acid	2.2	9.7	4.3	3.2
Histidine	1.8	9.2	6.0	7.6
Lysine	2.2	9.0	10.5	9.8
Tyrosine	2.2	9.1	10.1	5.7

Source: Table by Libretexts.

As anticipated, these kinds of compounds portray three inflection points in their titration curves, elucidated by the titrations of aspartic and arginine acid shown below. For majority of these compounds, there is possibility four charged species, in which one has overall charge.

Formulas for the given species are inscribed to the right of the titration curves, along with the pH at which each is projected to dominate. There is need of high amount of pH in a way to eradicate the final acidic proton from arginine which portrays the extremely great basicity of the guanidine moiety at the end of the side chain.

6.5.2. The Isoelectric Point

As it is noted above, the isoelectric point, referred as pI, is the pH of an aqueous solution of an amino acid or also known as peptide) at which on average the molecules have no net charge. In other context, it can be said that the groups which are positively charged are precisely balanced by the groups which are negatively charged. For amino acids which are simple in nature such as alanine, the pI is an average of the pK_a's of the ammonium (9.69) and carboxyl (2.34) groups.

Thus, the pI for alanine can be calculated as: (9.69+2.34)/2 = 6.02, the value which is determined experimentally. If additional basic or acidic groups are in existence as side-chain functions, the pI is the average of the pK_a's of the two acids which are identical in nature.

To find the similarity acids, these are divided into two categories. The first comprises of acids which are neutral in their protonated form (e.g., SH and CO_2H). The second comprises of acids which are positively charged in their protonated state (e.g., $-NH_3^+$). For the aspartic acid, the identical acids are the alpha-carboxyl function (pK_a = 2.1) and the side-chain carboxyl function (pK_a = 3.9), so pI = (2.1 + 3.9)/2 = 3.0. In case of arginine, the identical acids are the alpha-ammonium function (pK_a = 9.0) and guanidinium species on the side-chain (pK_a = 12.5), so the computed pI = (12.5 + 9.0)/2 = 10.75.

6.5.3. Other Natural Amino Acids

The 20 alpha-amino acids described above are the major components of proteins, their integration being regulated by the genetic code. There is existence of many other naturally occurring amino acids, and the structures of some of these are portrayed below. Some, such as hydroxyproline and hydroxylysine, are solely functionalized derivatives of earlier explained compound.

These two amino acids can be present only in collagen, which is a usual structural protein. Homocysteine and homoserine are greater homologs of their namesakes. The amino group that are in the category of beta-alanine has transferred to the end of the three-carbon chain.

It is a component which comes under the category of pantothenic acid, $HOCH_2C(CH_3)_2CH(OH)CONHCH_2CH_2CO_2H$, an essential nutrient and is adherent to the vitamin B complex. The gamma-amino homolog GABA is an antihypertensive agent and neurotransmitter inhibitor.

6.6. PROCESS OF RACEMIZATION

It is important to note that the optically active biphenyl derivatives can be racemized if at any time frame, the two aromatic rings proceed through a coplanar configuration by rotation about the central bond. This situation can be arisen with the help of heat either more or less easily, unless the 2,2-ortho substituent's are substantial in size.

The widely adopted method in which compounds with asymmetric carbon atoms are racemized is considered more intricate. One prospect for a tetrahedral chiral carbon is to combine it with four groups in order to become a chiral and planar without breaking any bonds.

With the help of theoretical calculations, it can be predicted that this would not be a process for chiral tetravalent carbon however, it is interesting to see that it will occur with chiral carbon and other chiral atoms that are connected with three groups:

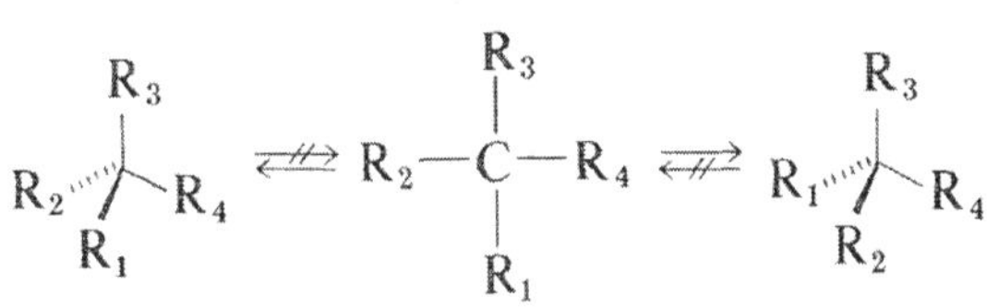

Optically, active carbonyl compounds in the category of–CHC=O in which there is asymmetric α-carbon, are racemized by both acids and bases, and are more likely to be associated with enolization. Construction of either the enolate or the enol anion will dismantle the asymmetry of a carbon so that, even if minimal amount of enol is in existence at any time frame, eventually racemized all of the compound will happen. However, the mechanism demands the requirement of the hydrogen and the need to place center of symmetry at this a carbon. Otherwise, acids, and bases would not be efficient and effective in catalyzing racemization.

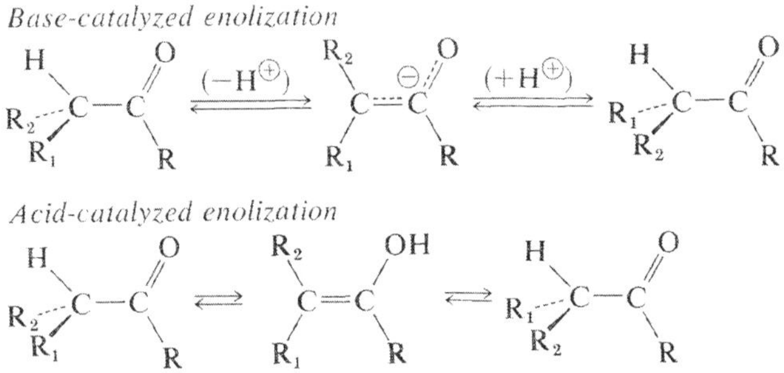

The racemization of an optically functioning subsidiary halide with the chiral carbon holding the halogen (for instance, 2-chlorobutane) may take place in solution and, generally, the better and frozen the ionizing the solvent is, the quicker the substance will racemized.

It can also be said that ionization of the halide through SN1 process likely to be responsible, and this irrefutably would be raised by polar solvents. There is significant evidence available that an alkyl carbocation once separated from its escort anion is planar; and, when this type of ion again come in contact with the anion, it has same chances of erecting it to L and D enantiomers:

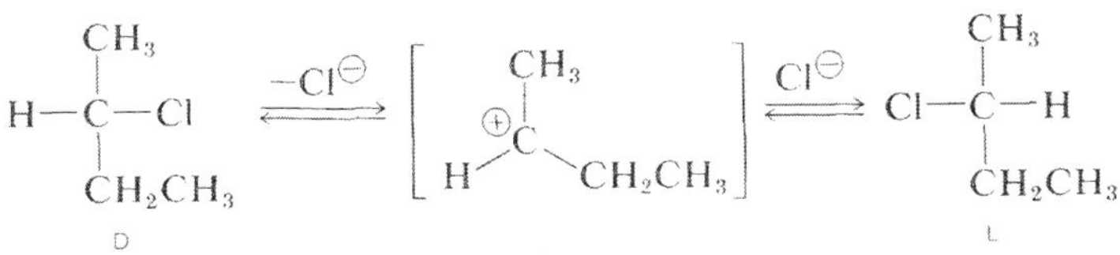

It is generally seen that optically working halides also can be racemized with the help of an SN2 mechanism. A resolution of active 2-chlorobutane in 2-propanone embracing solution of lithium chloride becomes racemic. The movement of the chloride of the halide by chloride ion trigger configuration at the atom experiencing interchanging activity. It resulted in introduction of a second substitution which regenerates the original enantiomer. Eventually, this repetitive process harness identical numbers of the L and D forms; the substance then is racemic:

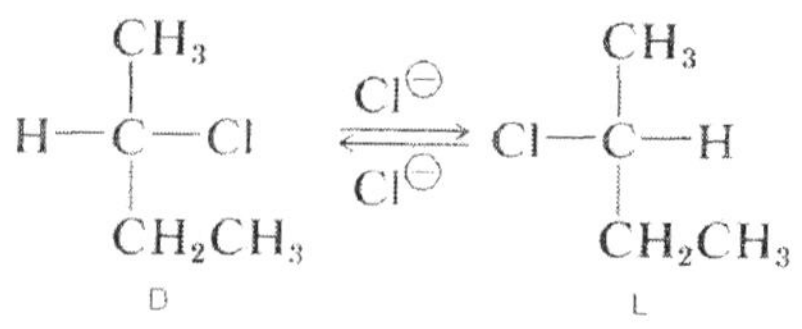

Asymmetric alcohols are in most of the cases are racemized by strong acids. Undoubtedly, it results in arising of an ionization and frequent encounter of the carbocation with water conducive to either enantiomer:

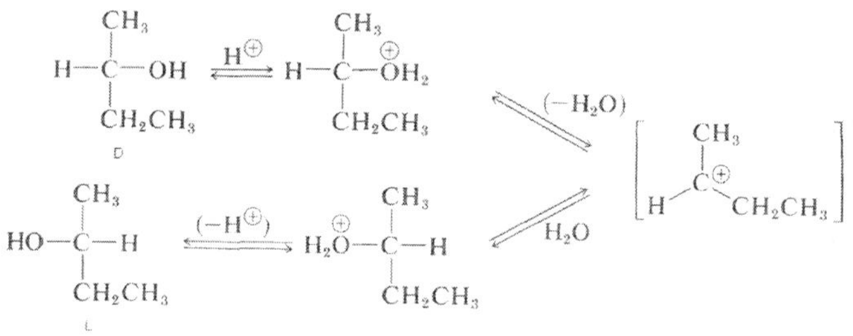

As compared to alcohols, halides, and carbonyl compounds, it is really a challenging task to racemize hydrocarbons. This is most likely the case when a compound having a quaternary asymmetric center, such as methyl ethyl propyl butyl methane, 28, that has no "handle" to find a way to transform into the asymmetric carbon to a symmetric condition with straight forward chemical methods:

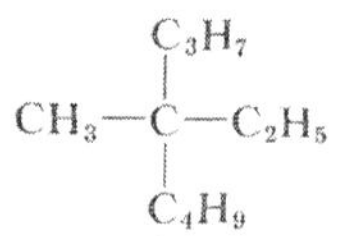

28

However, hydrocarbons which involves a hydrogen atom at the asymmetric carbon possess the ability to racemized if they can be transformed into either carbanions or carbocations. The facileness of carbanion-type racemization primarily relies on the acidity of the connected hydrogen and on the stability or consistency of stereochemical of the intermediate carbanion which is constructed. If the configuration of the carbanion intermediate inverts, then this will more likely result in racemization:

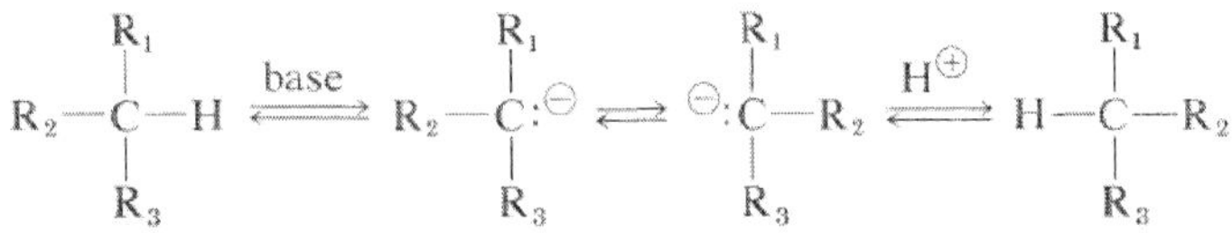

6.7. CONCLUSION

In the end, it is concluded that Stereoisomers are isomeric molecules which have the similar molecular constitution, but a completely distinct 3D spatial arrangement of the atoms. These are mainly categorized into diastereomers and enantiomers. Different types of enantiomers are generally associated with different receptors and possibly bring about distinctive reactions. Because of this, stereo selective separation methods are considered as the most important one in sectors such as food industry, pharmaceutical sector, and agricultural industry. Supercritical fluid chromatography and Liquid chromatography are the widely recognized tools in chiral separations.

There are various types of methods for each separation process. It necessitates the need to get an understanding about which methods is best in terms of efficiency and effectives.

It is generally seen that chiral compounds which are basically synthesized from a chiral starting materials and reagents are in most of the cases racemic in nature (for instance a 50:50 mixture of enantiomers). The process that is responsible for separation of racemates into their component enantiomers are known as resolution.

It was noticed that chromatography has freed resolutions from the constraint of reliance on crystallization as the approach on which diastereomer separation has been largely depended. Due to this reason, resolutions in general have shown more success rate at present compared with the past results. Not rarely oily covalent diastereomer mixtures ultimately form into crystals and their resolution then be function in a more customary way by getting the benefits of solubility differences.

As it is well known that enantiomers carry their own unique physical properties, such as melting point and solubility, resolution is a complicated task. While on the other hand, Diastereomers, have diverse physical properties, and this exclusive characteristic of diastereomers allow them to achieve resolution of racemates.

In various experiments, it was found that when a racemate is reacted with an enantiomerically pure chiral reagent, it gives a mixture of diastereomers that can be easily separated. When racemate's crystallization resulted in creation of a conglomerate then, by meaning, the element is said to be impulsively resolved. One can assume from this that during the process of racemate crystallization under equilibrium conditions, whether this occur immediately at a sluggish rate or at a rapid rate when persuaded with seed crystals, both enantiomers of the substance accumulated in identical amount as Enantiomorphs crystals.

In the chapter the meaning of triage is discussed which refers to the physical sorting of conglomerate crystals into segments whose outcomes are levorotatory and dextrorotatory (as computed by circular dichroism, polarimetry, or CD). It is widely believed that palpation of the crystal handedness can ramp up the development and this is achievable when there is well formation of crystals and this is endowed with hemihedral faces.

Stereoisomers which are not stable at a stereogenic center may be reacted which resulted in generation of stereo chemically modified forms. This can be prompted by an adjoining group that reduces consistency at the chiral center. In some of the cases, it was seen that enantiomer was 'racemize' irrevocably and reach the racemic mixture at the point of equilibrium. It can also 'enantiomerize,' or reversibly interconvert to the other enantiomer. Diastereomers can or 'epimerize' 'diastereomerize' by transformation of an in consistent stereogenic center.

REFERENCES

1. *(2007). Chemical Resolution of Enantiomers of 3,4-Dihydropyrimidin–2(1H)-Ones Using Chiral Auxiliary Approach.* [ebook] Available at: https://shodhganga.inflibnet.ac.in/bitstream/10603/10453/12/12_chapter%205.pdf (accessed on 18 August 2020).
2. Chemistry LibreTexts, (2009). *19.3: Separation or Resolution of Enantiomers.* [online] Available at: https://chem.libretexts.org/Bookshelves/Organic_Chemistry/Book%3A_Basic_Principles_of_Organic_Chemistry_(Roberts_and_Caserio)/19%3A_More_on_Stereochemistry/19.03%3A_Separation_or_Resolution_of_Enantiomers (accessed on 18 August 2020).
3. Chemistry LibreTexts, (2013). *19.11: Racemization.* [online] Available at: https://chem.libretexts.org/Bookshelves/Organic_Chemistry/Book%3A_Basic_Principles_of_Organic_Chemistry_(Roberts_and_Caserio)/19%3A_More_on_Stereochemistry/19.11%3A_Racemization (accessed on 18 August 2020).
4. Chemistry LibreTexts, (2014). *5.8: Racemic Mixtures and the Resolution of Enantiomers.* [online] Available at: https://chem.libretexts.org/Bookshelves/Organic_Chemistry/Map%3A_Organic_Chemistry_(McMurry)/05%3A_Stereochemistry_at_Tetrahedral_Centers/5.08%3A_Racemic_Mixtures_and_the_Resolution_of_Enantiomers (accessed on 18 August 2020).
5. Chemistry LibreTexts, (2019). *19.4: Enantiomeric Purity.* [online] Available at: https://chem.libretexts.org/Bookshelves/Organic_Chemistry/Book%3A_Basic_Principles_of_Organic_Chemistry_(Roberts_and_Caserio)/19%3A_More_on_Stereochemistry/19.04%3A_Enantiomeric_Purity (accessed on 18 August 2020).
6. Chemistry LibreTexts, (2019). *5.2: Configurational Isomers.* [online] Available at: https://chem.libretexts.org/Bookshelves/Organic_Chemistry/Book%3A_Basic_Principles_of_Organic_Chemistry_(Roberts_and_Caserio)/05%3A_Stereoisomerism_of_Organic_Molecules/5.02%3A_Configurational_Isomers (accessed on 18 August 2020).
7. Chemistry LibreTexts, (2020). *23.7: The Resolution of Racemic Mixtures of Amino Acids.* [online] Available at: https://chem.libretexts.org/Bookshelves/Organic_Chemistry/Map%3A_Organic_Chemistry_

(Bruice)/23%3A_The_Organic_Chemistry_of_Amino_Acids%2C_Peptides%2C_and_Proteins/23.07%3A_The_Resolution_of_Racemic_Mixtures_of_Amino_Acids (accessed on 18 August 2020).

8. De Torres, T., (1999). *Racemization - An Overview | Sciencedirect Topics*. [online] Sciencedirect.com. Available at: https://www.sciencedirect.com/topics/earth-and-planetary-sciences/racemization (accessed on 18 August 2020).
9. Eliel, E., Wilen, S., & Mander, L., (1994). Stereochemistry of Organic Compounds. New York: John Wiley & Sons.
10. Faigl, F., Fogassy, E., Nogradi, M., & Palovics, E., (2019). *Separation of Non Racemic Mixtures of Enantiomers: An Essential Part of Optical Resolution*. [online] ResearchGate. Available at: https://www.researchgate.net/publication/41467318_Separation_of_non-racemic_mixtures_of_enantiomers_An_essential_part_of_optical_resolution (accessed on 18 August 2020).
11. Galea, C., Vander, H. Y., & Mangelings, D., (2020). *Chapter 12 - Separation of Stereoisomers*. [online] www.sciencedirect.com. Available at: https://www.sciencedirect.com/science/article/pii/B9780128092071000124 (accessed on 18 August 2020).
12. Hou, G., Zi, G., Li, M., Wang, R., Han, S., & Kong, D., (2014). *Kinetic Resolution of Racemic 2-Substituted 1,2-Dihydroquinolines Via Asymmetric Cu-Catalyzed Borylation*. [online] https://pubs.rsc.org. Available at: https://pubs.rsc.org/en/content/articlelanding/2017/sc/c7sc01556a#!divAbstract (accessed on 18 August 2020).
13. Kanamori, Y., Watanabe, M., Kawauchi, K., chen, Y., Yanagishita, H., & Hirata, H., (2004). *Kinetic Resolution of Enantiomers in Racemic and Enantiomerically Enriched 2-Alkanols by Pseudomonas Cepacia Lipase Catalyzed Transesterification with Isopropenyl Acetate in Organic Solvent*. [ebook] Available at: https://www.jstage.jst.go.jp/article/jos/54/1/54_1_21/_pdf (accessed on 18 August 2020).
14. Leek, H., & Andersson, S., (2017). Preparative scale resolution of enantiomers enables accelerated drug discovery and development. *Molecules*, *22*(1), 158.
15. Merriam-webster.com. (2015). *Definition of Racemization*. [online] Available at: https://www.merriam-webster.com/dictionary/racemization (accessed on 18 August 2020).

16. Reist, M., Testa, B., Carrupt, P., Jung, M., & Schurig, V., (2020). *Racemization, Enantiomerization, Diastereomerization, and Epimerization: Their Meaning and Pharmacological Significance.* [online] onlinelibrary.wiley.com. Available at: https://onlinelibrary.wiley.com/doi/10.1002/chir.530070603 (accessed on 18 August 2020).
17. Roy, S., Chen, K., Gurubrahamam, R., & Chen, K., (2014). Organocatalytic kinetic resolution of racemic secondary nitroallylic alcohols combined with simultaneous desymmetrization of prochiral cyclic anhydrides. *The Journal of Organic Chemistry*, *79*(19), 8955–8959.
18. S. B., (2015). *Racemization in Organic Chemistry: Definition, Mechanism and Example.* [online] Study.com. Available at: https://study.com/academy/lesson/racemization-in-organic-chemistry-definition-mechanism-example.html (accessed on 18 August 2020).
19. Tcichemicals.com. (2013). The Kinetic Resolution of Racemic Alcohols and Racemic Carboxylic Acids. [online] Available at: https://www.tcichemicals.com/en/in/support-download/tcimail/application/141–18.html (accessed on 18 August 2020).

Chapter 7

Stereochemistry of Chemical Reaction

CONTENTS

In the chemical reactions, it is very important to understand the spatial arrangement of the atoms. This chapter explains the basics of the relation between the stereochemistry and the chemical reactions. This chapter introduces to the importance of stereochemistry and the importance of organic chemistry from the aspect of the chemical reactions. The concept of the optical isomers along with the set-up of the polarimeter is discussed in this chapter. This chapter also sheds light on the asymmetry of the molecule.

The important aspect of the configuration of the stereochemistry is discussed in this chapter. Along with the mentions of configuration, the retention and the inversion of the configuration of the stereochemistry is discussed in this chapter. After the discussion of the configuration, the process of the racemization is explained in the chapter. Various types of stereoisomers and different aspects of stereoisomerism are discussed in this chapter.

The importance of the stereochemistry and the important aspects of the absolute configuration are discussed in this chapter.

7.1. INTRODUCTION

Stereochemistry is the branch of chemistry that study of the arrangement of atoms in space. Stereoisomers are defined as the compounds that are having the same molecular formula but these compounds that differ in the spatial arrangement (that is the arrangement in space). There are two main categories of stereoisomers:

- Geometric isomers; and
- Optical isomers.

The proper understanding of optical isomers and their involvement in these chemical reactions is very important.

Stereochemistry is defined as the study of the relative arrangement of atoms or groups of the atoms in a molecule in three-dimensional space. Stereochemistry is defined as a subdiscipline of chemistry that includes the study of the relative spatial arrangement of atoms. This forms the structure of molecules as well as the manipulation of the molecules.

The study of stereochemistry pays attention on the stereoisomers. By the definition, these stereoisomers have the same molecular formula as well as the sequence of bonded atoms (constitution), but these stereoisomers vary in the three-dimensional orientations of the atoms in space. It is also called ad 3D chemistry because of this reason. In this, the prefix "stereo-" means "three-dimensionality." The study of chiral molecules is the important branch of stereochemistry.

This branch of chemistry that is stereochemistry spans the entire spectrum of inorganic, organic, biological, physical, and mainly the supramolecular chemistry. Stereochemistry consists of the methods for determining as well as describing all these relationships; the impact on the biological properties or physical properties these relationships impart upon the molecules in question, and the ways in which these associations have impact on the reactivity of the molecules in question. It has to be noted that stereochemistry is a vital issue in any synthesis.

There are some studies that explain two important points. First, at a CC bond, the disconnection should be done. Here, one of the atoms of carbon is a stereogenic center. The disconnection of a bond that is away from the stereogenic center generally give rise to a less efficient as well as less desirable retro synthesis. Often, it is more tough.

The second important point deals with search engines. It has been observed that searching exact structures with the entire combination of the stereochemistry may return no hits. On the other hand, the similar kind of search for the racemic structure may return many hits or it may return at least related structures that can help with the process of planning.

The important lesson is that an individual should not restrict the search to the structure that includes "wedges" and "dashes," but it should also search by the use of the racemic structure. Certainly, it may be more beneficial to start the search with the racemic compound and make use of that information in order to guide any kind of search with the enantiopure compound.

Stereogenic carbon is the central atom. This is because it consists of four distinct groups. These groups are Br, Me, Et, and H. A chiral object is defined as an object that has a non-super imposable mirror image. Here, the two versions of 2-bromobutane on both the side of the mirror are enantiomers.

It should be noted that there exists no such thing as a "chiral carbon"; rather, it is said that it is the stereogenic carbons that cause the entire molecule to be chiral. Also, it should be noticed that it is possible for a molecule to have stereogenic carbons and yet the molecule can be a chiral. The two molecules that are mentioned above have the similar properties in every possible way.

7.2. IMPORTANCE OF STEREOCHEMISTRY

The main aim of stereochemistry is to described the natural phenomena of spatial arrangements of the organic molecules. Stereocenter is defined as any atom in a molecule, this atom is attached to four distinct atoms. It is also called as chiral center. For instance, a molecule that are having three stereocenters would give rise to a molecule that is having 8 stereoisomer's.

It has been observed that chemists can work out the relationships that is between these different molecules that are formed of the same atoms. For this, chemists make use of stereochemistry. In addition to it, the chemists can also study the impact on the physical properties or biological properties of these molecules.

7.2.1. Importance of Organic chemistry

Organic chemistry is the branch of chemistry that is very important. This is because it is the study of life and the study of all of the chemical reactions that are associated to the life. There are various careers that apply an understanding of organic chemistry, like the dentists, doctors, veterinarians, chemical engineers, pharmacologists, and chemists.

In addition to it, organic chemistry also plays a very import part in the day-to-day life because clothes, ink, rubber, soap, medicines, food, paper, perfumes etc. are indispensable to the people for good living. It has to be noted that Organic compounds are the vital constituents of various products for example, food, medicine, paint, explosive, pesticide, plastic, explosive, petrochemical, pesticide etc.

The branch of chemistry, organic chemistry is a highly creative science in which the chemists make new molecules and discover the properties of the compounds that are existing. It is considered as the most popular field of study for PhD Chemists and ACS chemists. Human beings are surrounded by countless number of organic compounds.

7.3. OPTICAL ISOMERS

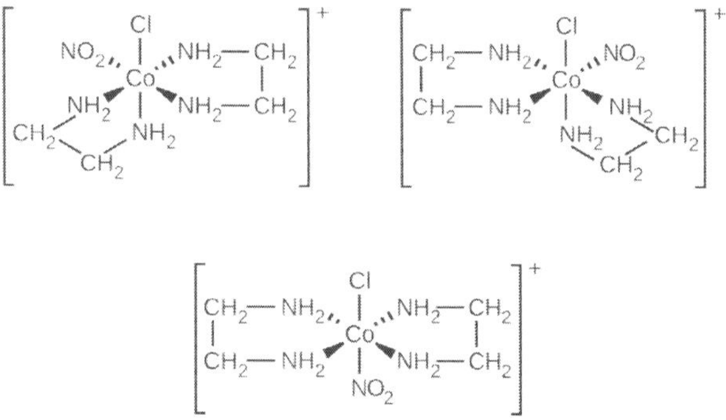

Optical isomers

Figure 7.1. Optical isomers have the same molecular formula but have different passage of plane polarized light as they are rotated.

Source: Image by Wikimedia commons.

> **Optical isomers are defined as the compounds that are having the same molecular formula, but these compounds vary in the way they rotate a plane polarized light**

Plane polarized light is formed when an ordinary light is passed through a Nicol prism, thus, it causes diffraction and results in the formation of a single wavelength light. Plane polarized light is the light of the single wavelength (Figure 7.1).

7.3.1. Optically Active Compounds and Their "Magical Powers"

Some of the researchers found that some optically active reactants possess some kind of 'magical powers.' This might prove to be helpful to some and might appear rather silly to the others. But there are scientists that believed in some kind of magical power was rested in the optically active compounds.

This was majorly because of the observations that these scientists made during the SN2 reaction. In the SN2 reaction, the molecule retains its chirality throughout the reaction, thus retaining its magical powers and remains optically active.

On the other hand, in a SN1 reaction, the molecule loses its chirality in the middle of the reaction. This made the scientists believe that the magical powers of the molecule are dropped. Now, as the optical activity of the molecule is lost, any product that was chiral at the start of the reaction will now become racemic Now, the only way to get back the magical powers (the optical activity) is to bring another magical molecule (chiral reactant) in contact with the achiral molecule.

7.4. ASYMMETRY OF A MOLECULE

Whether the molecule will be chiral or achiral is determined by the asymmetry of a molecule. There is the availability of various objects or compounds that are non-super imposable. These kinds of objects are called as chiral objects, or they can also be called as chiral compound.

However, it has been observed that there are various compounds or objects, like a sphere, that is the same with its mirror image as well as it is superimposable. This kind of compounds or objects is known as achiral object. Thus, the chirality of the molecule will be determined by the asymmetry of a carbon atom in a particular molecule

Why there is the need to study chirality? The reason is that it will help in studying about enantiomers. Enantiomers are nothing but these are the stereoisomers that are not able to form super imposable mirror images of each other. In addition to it, there are some physical properties of enantiomers. These properties involve the boiling point, index of refraction etc. It has been observed that these compounds differ in the manner of rotation of a plane polarized light (that is dextro-or levo).

Though, if the specific composition of a mixture includes the same amount of both the kinds of enantiomers, zero is the net optical rotation. This is because dextro-and levo-counter each other. This kind of mixture is defined as a racemic mixture and the process is known known as racemization.

7.5. IMPORTANCE OF CONFIGURATION IN STEREOCHEMISTRY

It has been observed that there are different kinds of spatial arrangement in an organic compound. There are various research papers and journals that explains about the spatial arrangements of an organic compound with respect to stereochemistry.

7.5.1. Retention of Configuration

Retention of configuration is defined as the ability of an asymmetric carbon to keep the spatial arrangement of the substituent's or groups intact at the time of a chemical reaction. This can only happen when the bond of carbon stereo center continue to be intact. Thus, a substituent will be substituted by the other substitute without troubling the entire structure. Therefore, the relative arrangement of the reactant as well as the product can be connected.

7.5.2. Inversion of Configuration

If incase an asymmetric carbon atom is unable to retain its spatial arrangement at the time of chemical reaction, then it is called as inversion. It has to be noted that during SN2 reactions, an inversion takes place. This is because the C-X bond of an alkyl halide is attacked by the nucleophile by the help of backside method. The configuration of the molecule gets inverted as a result of this. Thus, in an SN2 reaction, the optically active alkyl halides give rise to the inversion of configuration.

7.6. FORMATION AND PROPERTIES OF RACEMIC MIXTURES

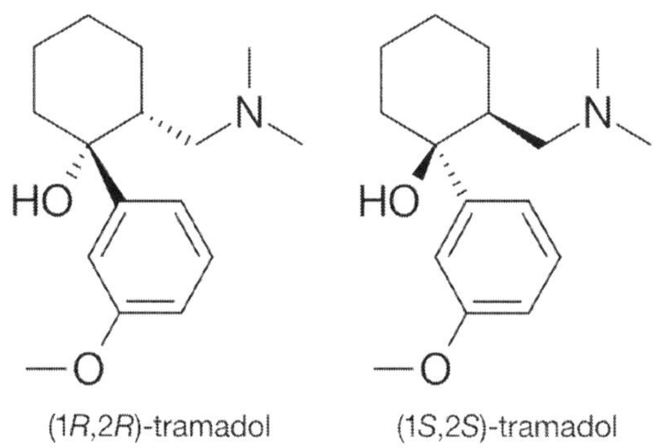

Figure 7.2. Racemic mixtures are the combination of two different enantiomers.

Source: Image by Wikipedia.

A racemic blend is a 50:50 blend of two enantiomers. Since they are identical representations, every enantiomer pivot plane-enraptured light in an equivalent yet inverse way and is optically dormant. In the event that the enantiomers are isolated, the blend is said to have been settled. A typical test in the research center part of early on natural science includes the goals of a racemic blend (Figure 7.2).

The subjective biochemical results of chirality are shown by the utilization, during the 1950s, of the medication Thalidomide, a sedative given to pregnant ladies to mitigate morning ailment. It was later understood that while the (+)-form of the particle, was a safe and successful narcotic, the (−)-form was a functioning teratogen. The medication caused various birth abnormalities during the early stages when taken in the beginning periods of pregnancy since it contained a blend of the two structures.

As noted, before, chiral mixes blended from achiral beginning materials and reagents are for the most part racemic (which is a 50:50 blend of enantiomers). Division of racemates into their part enantiomers is a procedure called goals. Since enantiomers have indistinguishable physical properties, for example, solvency, and dissolving point, goals are very troublesome. Diastereomers, then again, have distinctive physical properties, and this reality is utilized to accomplish goals of racemates.

The response of a racemate with an enantiomerically unadulterated chiral reagent gives a blend of diastereomers, which can be isolated. For instance, if a racemic blend of a chiral liquor is responded with an enantiomerically unadulterated carboxylic corrosive, the outcome is a blend of diastereomers: right now, the unadulterated (R) enantiomer of the corrosive was utilized, the item is a blend of (R-R) and (R-S) diastereomeric esters, which can, in principle, be isolated by their distinctive physical properties. Resulting hydrolysis of each isolated ester will yield the 'settled' (enantiomerically unadulterated).

As noted, before, chiral mixes combined from achiral beginning materials and reagents are by and large racemic (e.g., a 50:50 blend of enantiomers). Partition of racemates into their segment enantiomers is a procedure called goals. Since enantiomers have indistinguishable physical properties, for example, dissolvability, and softening points, goals are very troublesome.

Diastereomers, then again, have diverse physical properties, and this reality is utilized to accomplish goals of racemates. The response of a race mate with an enantiomerically unadulterated chiral reagent gives a blend of diastereomers, which can be isolated. Turning around the primary response at that point prompts the isolated enantiomers in addition to the recouped reagent.

Numerous sorts of synthetic and physical responses, including salt development, might be utilized to accomplish the diastereomeric intermediates required for partition. The above figure represents this general guideline by demonstrating how a nut having a right-gave string (R) could fill in as a "reagent" to segregate and separate a blend of both ways gave electrical discharges size and weight.

Just the two right-gave accomplices can associate to give a completely strung middle of the road, so division is genuinely straightforward. The settling moiety, for example, the nut, is then expelled, leaving the fasteners isolated into their privilege and left-gave structures. Substance responses of enantiomers are typically not all that significantly unique, yet a commonsense differentiation is in any case conceivable.

Since the physical properties of enantiomers are indistinguishable, they only from time to time can be isolated by basic physical techniques, for example, fragmentary crystallization or refining. It is just affected by another chiral substance that enantiomers carry on in an unexpected way, and practically all techniques for goals of enantiomers depend on this reality.

7.6.1. Chiral Amines as Resolving Agents and Resolution of Racemic Acids

The most ordinarily utilized strategy for isolating enantiomers is to change over them to a blend of diastereomers that will have diverse physical properties: dissolving point, breaking point, dissolvability, etc. For instance, in the event that you have a racemic or D, L blend of enantiomers of a corrosive and convert this to a grain of salt with a chiral base having the D arrangement, the salt will be a blend of two diastereomers: (D corrosive. D base) and (L corrosive. D base).

These diastereomeric salts are not indistinguishable and they are not perfect representations. In this way, they will vary somewhat in their physical properties, and a detachment by physical strategies, for example, crystallization, might be conceivable. On the off chance that the diastereomeric salts can be totally isolated, the corrosive recovered from each salt will be either solely the D or the L enantiomer (Figure 7.3).

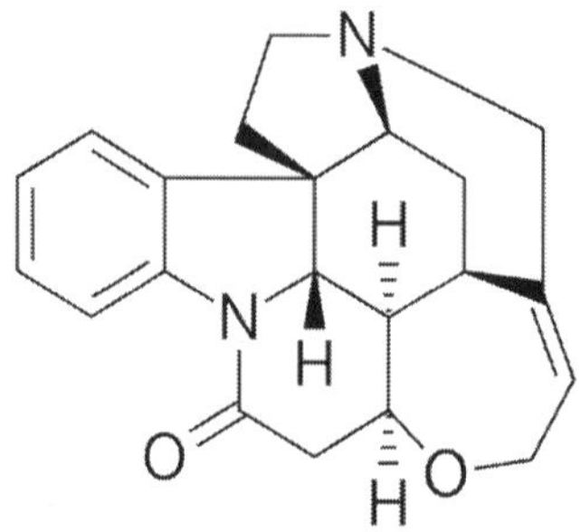

Figure 7.3. Strychnine is a chiral acid that finds its usage in almost every stereochemical reaction.

Source: Image by Wikipedia.

The goals of chiral acids through the arrangement of diastereomeric salts requires sufficient supplies of reasonable chiral bases. Brucine, strychnine, and quinine every now and again are utilized for this reason since they are promptly accessible, normally happening chiral bases. Less difficult amines of engineered source, for example, 2-amino-1-butanol, amphetamine, and 1-phenylethanamine, likewise can be utilized, yet first, they should be settled themselves.

7.6.2. Resolution of Racemic Bases

Chiral acids, for example, (+)-tartaric corrosive, (–)-malic corrosive, (–)-mandelic corrosive, and (+)-camphor–10-sulfonic corrosive, are utilized for the goals of a racemic base.

The standard is equivalent to the goals of a racemic corrosive with a chiral base, and the decision of corrosive will depend both on the simplicity of detachment of the diastereomeric salts and, obviously, on the accessibility of the corrosive for the size of the goals in question. Goals strategies for this sort can be dull, in light of the fact that various recrystallizations in various solvents might be important to logically advance the precious stones in the less-dissolvable diastereomer.

To decide when the goals are finished, the blend of diastereomers is recrystallized until there is no further change in the deliberate optical revolution of the precious stones. At this stage, it is trusted that the crystalline salt is an unadulterated diastereomer from which one unadulterated enantiomer can be recuperated.

The optical revolution of this enantiomer will be the greatest worth on the off chance that it is "optically" unadulterated in light of the fact that any measure of the other enantiomer could just diminish the greatness of the deliberate pivot.

7.6.3. Resolution of Racemic Alcohols

To determine a racemic liquor, a chiral corrosive can be utilized to change over the liquor to a blend of diastereomeric esters. This isn't as commonly valuable as may be suspected on the grounds that esters will, in general, be fluids except if they are extremely high-molecular weight mixes. In the event that the diastereomeric esters are not crystalline, they should be isolated by some other strategy than partial crystallization (e.g., by chromatography strategies). Two chiral acids which play very important role in resolving agents for alcohols are explained as follows:

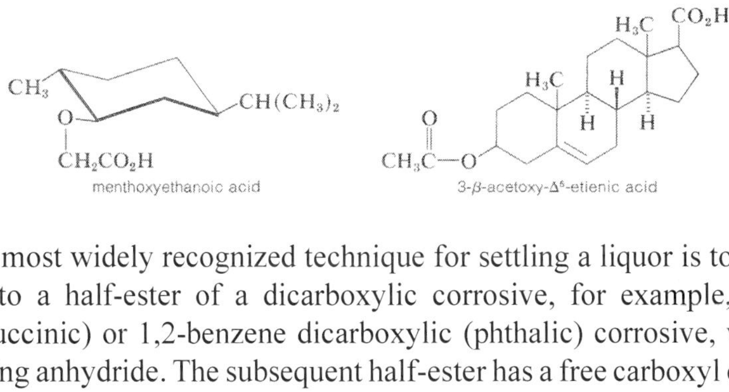

menthoxyethanoic acid

3-β-acetoxy-Δ5-etienic acid

The most widely recognized technique for settling a liquor is to change over it to a half-ester of a dicarboxylic corrosive, for example, butane dioic (succinic) or 1,2-benzene dicarboxylic (phthalic) corrosive, with the comparing anhydride. The subsequent half-ester has a free carboxyl capacity and may then be resolvable with a chiral base, generally brucine.

7.6.4. Other Methods of Resolution

One of the significant objectives in the field of natural science is the improvement of reagents with the property of "chiral acknowledgment" to such an extent that they can impact a spotless partition of enantiomers in a single activity without crushing both of the enantiomers. It might not have been accomplished to the perfection, yet it may not be far later on.

Chromatographic strategies (identified with TLC and GC), whereby the stationary stage is a chiral reagent that adsorbs one enantiomer more unequivocally than the other, have been utilized to determine racemic mixes, however such goals only here and there have prompted both unadulterated enantiomers on a preparative scale. Different strategies, called motor goals, are incredible when relevant, exploiting contrasts in response paces of enantiomers with chiral reagents.

One enantiomer may respond more quickly than the other, in this way leaving an overabundance of the other enantiomer behind. For instance, racemic tartaric corrosive can be settled with the guide of certain penicillin shapes that devour the dextrorotatory enantiomer quicker than the levorotatory enantiomer. Subsequently, practically unadulterated (–)-tartaric corrosive can be recuperated from the blend:

(±)-tartaric corrosive + form → (–)-tartaric corrosive + more shape

An impediment of goals of this sort is that the more receptive enantiomer, as a rule, isn't recoverable from the response blend.

The crystallization methodology utilized by Pasteur for his traditional goals of (±)-tartaric corrosive has been effective just in not very many cases. This method relies upon the development of individual gems of every enantiomer. Along these lines if the crystallization of sodium ammonium tartrate is done beneath 27″, the typical racemate salt doesn't shape; a blend of precious stones of the (+) and (–) salts frames. The two various types of gems, which are connected as an item to its perfect representation, can be isolated physically with the guide of a magnifying instrument and in this way might be changed over to the tartaric corrosive enantiomers by solid corrosive.

A minor departure from this technique for goals is the seeding of an immersed arrangement of a racemic blend with gems of one unadulterated enantiomer in the desire for causing crystallization of simply that one enantiomer, along these lines leaving the other in an arrangement. Shockingly, not very many down to earth goals have been accomplished right now.

In any event, when a fruitful is accomplished, some noteworthy issues remain. For example, the goals itself doesn't give data on the real design of the (+) or (–) enantiomer. Likewise, it is beyond the realm of imagination to expect to tell the enantiomeric immaculateness (optical virtue) of the settled enantiomers without extra data.

By the consideration of the two-step process that takes place in SN1 reactions, it will be noted that in the first step in SN1 reaction, planar carbocation generation and in the second step attack of a nucleophile takes place. The carbocation is planar that permits the nucleophile in the second step to attack the atom of the carbon either from above or below. In turn, this results in both configurations-retention, and inversion.

In addition to it, there is also a probability that 50% of the mixture will have the retention of configuration and 50% will have inversion of configuration. In the example that is mentioned below, both inversion of configuration and retention of configuration have taken place with regard to the actual reactant. Retention configuration is ((−) Butan-2-ol) while the inversion configuration is ((+)Butan-2-ol).

Therefore, there is the formation of three different configurations that includes retention of configuration, inversion of configuration, and racemization, as the result of a reaction of an asymmetric carbon.

7.7. STEREOISOMERISM AND DIFFERENT TYPES OF ISOMERS

Table 7.1. Types of Stereoisomers

Atropisomerism	Atropisomerism is defined as the property of any molecule or an object of being non-superimposable on its mirror image.
Cis-trans isomerism	Cis-trans isomerism shares the same atoms that are joined to each another in the similar way, but they have a different configuration.
Conformational isomerism	Conformational isomerism is defined as a kind of stereoisomerism in which isomers can only be converted by formally single bond rotations.
Diastereomers	Diastereomers are defined as optically active isomers which are not enantiomers.
Enantiomers	An enantiomer is one of a pair of optical isomers, the structures of which are not superimposable on their mirror images.

Source: Table from Byju's.com.

Stereoisomerism is referred to "the isomerism that the result of the non-similar arrangements of atoms or it is formed by the functional groups belonging to an atom in space." It has to be noted that these kinds of isomers have the same constitutions, but these isomers have different geometric arrangements of atoms. In addition to it, stereoisomers can be categorized broadly into two kinds (Table 7.1).

- Enantiomers; and
- diastereomers.

7.7.1. Enantiomers

a. The type of isomerism is known as enantiomerism when two of the isomers are mirror images of each other. These kinds of isomers are known as enantiomers.
b. Enantiomers are defined as the stable and isolable compounds that varies in their spatial arrangements in 3-D space.
c. Enantiomers are the isomers that usually exist as discreet pairs.
d. All the properties of enantiomers are the same. Though, the interaction of these enantiomers with a plane of polarized light can be different.
e. The direction in which these enantiomers rotate the plane-polarized light is not the same. It means that if one rotates in the right direction than the other one rotates towards the left direction.

7.7.2. Diastereomers

a. The isomers are known as diastereomers when two isomers do not behave as mirror images of each other.
b. A molecule that is having 'n' number of asymmetric atoms of carbon atoms can have up to '2n' diastereomers.
c. Diastereomers are referred to as epimers when two diastereomers vary at only one stereocenter.
d. These kinds of isomers have different physical properties and also differ in chemical reactivity.

7.8. CHIRALITY AND DIFFERENT SUBSTITUENTS ON TETRAHEDRAL CARBON

It has to be noted that when a tetrahedral carbon can be changed to a chiral center by just bringing the change in only one of its attached groups, it is described as a 'prochiral' center. The actual example that is shown below is the reduced form of a molecule which is known as nicotinamide adenine dinucleotide (NADH), a vital participant in various biochemical oxidation reactions or reduction reactions (Figure 7.4).

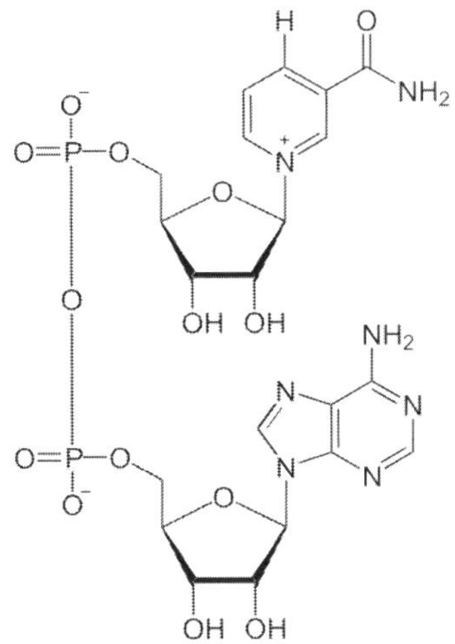

Figure 7.4. NADH has proven to be an important reactant in the biochemical reactions.

Source: Image by Wikipedia.

It has to be noted that in case either one of the indicated hydrogens on NADH is changed to a deuterium (that is the 2H isotope of hydrogen) then the carbon will become a chiral center. Prochirality is considered as a vital concept in biological chemistry.

This is because enzymes can differentiate between the two 'identical' groups that are bound to a prochiral carbon center because of the fact that they occupy distinct regions in three-dimensional space. For instance, in the following reaction, which is an important step in the oxidation of fatty acids, it is mainly HA and HD that are lost. On the other hand, HB, and HC continue to be in the resulting conjugated alkene.

The hydrogens, which are prochiral in nature, on C2 and C3 of the fatty acid can be designated in accordance to a variation on the R's system. To make it clear and evident, it has been observed that everyone looks at a much simpler molecule (that is ethanol) in order to defined this system.

In order to name the two prochiral hydrogens on ethanol, again, one is required to take part in a thought experiment. If an individual, in his or her imagination, was to randomly change HB to a deuterium then the molecule would now be chiral. In this, the stereocenter would have the S configuration (it has to be noted that D has a higher priority as compared to that of H).

Because of this reason, one can refer to HB as the pro-S hydrogen of ethanol, and name it as HS. On the other hand, if HA is changed to D and HB is left as a hydrogen, the configuration of the molecule becomes R. So, HA can be referred as the pro-R hydrogen of ethanol, and it can be labelled as HR.

If one looks back to all the examples of the fatty acid, he or she may see that it is exactly the pro-R hydrogens on carbons 2 and 3 that are lost in the reaction.

The prochiral hydrogens can be defined either as enantiotropic or diastereotopic. If a deuterium is used to replace either the HR or the HS, which are attached to the ethanol, then the two resulting molecules would be enantiomers. This is so because there are no other stereocenters. Therefore, these two hydrogens are known as enantiotropic.

Although, something different is seen in glyceraldehyde-3-phosphate (GAP).

The two resulting molecules will be diastereomers if either HR or HS is substituted by a deuterium. Therefore, in this molecule, HR, and HS are known as diastereotopic hydrogens. The significance of the difference between enantiotropic and diastereotopic groups will become apparent when one acquires knowledge about the analytical technique that is known as nuclear magnetic resonance (NMR).

Two atoms of hydrogens that are on the same carbon of a substituted ring structure can be diastereotopic. This can be determined by conducting the same thought experiment as it was discussed above. In the example that is mentioned below, the diastereotopic hydrogens that are indicated are either on the same side of the ring or they are on the opposite side of the ring that is relative to the hydroxyl group.

Finally, the atoms of hydrogens that are completely similar and can be named neither enantiotropic nor diastereotopic are known as homotopic. If in case a homotopic hydrogen is substituted by deuterium, there is no creation of a chiral center. It has to be noted that the three atoms of hydrogen that are on the methyl (CH_3) group of ethanol (and on any methyl group) are homotopic in nature.

Even to an enzyme, all three of these hydrogens will look the same.

7.8.1. Reactions at a Single Chiral Carbon (chiral → chiral)

During the reaction, if a bond to a chiral carbon breaks, there are various possibilities:

a. The old bond is broken down and new bond is formed. This formation takes place in a single step, or in an ordered way. In this case, the stereochemistry will either be retained or it will be inverted.

b. Bond is broken and there is the formation of new bond. It takes place in two or more than two steps (that is, via intermediates), or it is formed in a less ordered way. In this situation, there might be less distinction in the stereochemistry, and as a result of it, racemic mixture will be formed (this mixture is 50:50 mixture of enantiomers).

Here, the alkyl halide is attacked by iodide (I^-) from the back and it makes a product. In this product, the chiral center has been inverted. It has been observed that there is no scrambling of the stereochemistry, and the product would remain completely optically active. This is because the process takes place all in just a single step.

The alkyl halide is attacked again by the iodide ion. But in this situation, the reaction occurs in two steps. In the reaction, just after the first step an achiral intermediate is made, and as a result of this, a racemic (50:50 mixture of enantiomers) mixture of enantiomers for the chiral product is formed.

7.8.2. Reactions That Form a New Chiral Center

If the achiral carbon is the initial or the starting material, then there will be the formation of racemic mixture of the product enantiomers. This will be the case until some other chiral agent gets involved in the reaction. The SN1 reaction above gives a racemic product because of the same reason. Also, it has to be observed that there is an equal chance of attacking the achiral starting material from any of the side. Therefore, Achiral →Chiral will form the racemic product.

On the other hand, if chiral is already the starting material, and there is the introduction of a new chiral center. This situation is more complex. This is because there might be the production of a mixture of diastereomers. The mixture of diastereomers have varying stabilities.

7.9. IMPORTANCE OF RELATIVE STEREO DESCRIPTORS

It must be noted that even this chiral molecular entity (or a group) possesses an entirely different and separate stereochemical description. For example, the chiral molecular entity (or the groups) can be denoted by R or S; and α (alpha), β (beta). Many researchers have carefully observed that these different stereo descriptors are put to use in various ways.

1. Relative stereo descriptors are used to describe the configuration at the anomeric carbon in carbohydrate nomenclature. This is done by associating the anomeric carbon to the anomeric reference atom. It has been noted that in simple cases, the anomeric reference atom and the reference atom of the configuration, both are same. Thus, in a-r>-glucopyranose, the configurational reference atom, which is C–5 and the OH atom at C–1 is on the same side as the OH at the C–5 as observed in the given Fischer projection.
2. There are some Relative stereo descriptors that are used by Chemical Abstracts (CA) Service. These relative stereo descriptors are used to explain and describe the configuration of a cyclic molecule. These relative stereo descriptors are also used to describe the suitable polycyclic systems. Most of the compounds, that use the relative stereo descriptors, usually have a number of stereogenic centers. Here, the side of the reference plane is referred to as the side on which the substituent with CIP priority lies at the lowest numbered stereogenic center.
3. It must be noted that the absolute stereo descriptors were originally formulated for the naming the steroids. However, according to this point of view, this can only be meaningful if there is an absolute configuration and orientation of the structure. This structure and orientation must be agreed so that the plane of orientation must be defined and which way up the molecule is represented. Substituent's above the plane of the steroid are described as p and shown as a solid line (-or-), those below the plane are described as a and shown by a broken line (11111111 or -----).

There have been various documents that state the extension of this system to tetrapyrroles. Also, there have been many instances displaying that this has been used elsewhere. In this system, a prefix is used to depict or indicate a molecule. This molecule possesses two (or more) chiral elements.

Also, this molecule is present as a mixture of the two racemic diastereomers that are present in the unspecified proportions. For example, the dipeptide formed from L-alanine and DL-leucine is L-alanyl-ambo-leucine.

7.10. ANGLE STRAIN

Angle strain can be defined as the strain that is observed due to a deviation in the bond angle from the "standard" or "normal" values. The term angle strain is usually put to use in the context of the nonaromatic cyclic compounds. It has been noticed that in the nonaromatic cyclic compounds have the internal angles that vary from the standard value of tetrahedral angle of 109" 28'. Thus, the angles in the nonaromatic cyclic compounds experience angle strain. This strain is also known as 'Baeyer strain.'

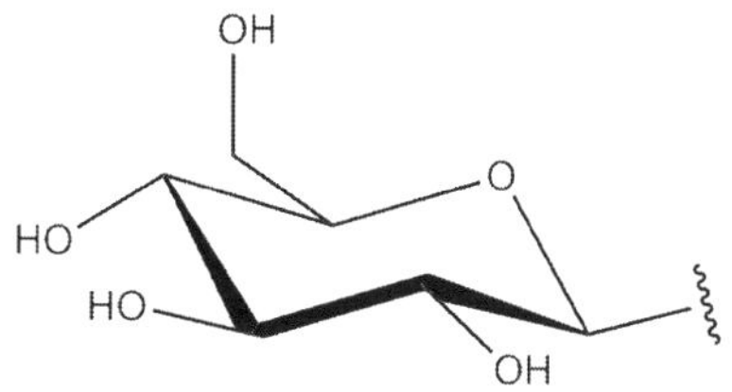

Figure 7.5. Beta-D-glucopyranosyl shows anomeric effect.

Source: Image by Wikimedia Commons.

Anomeric Effect can be explained originally as the result of the thermodynamic preference of the polar groups, which are bonded to C–1, to gain an axial position in the compound. The C–1, to which the polar groups are bonded, is the anomeric carbon of a glucopyranosyl derivative. It is due to anomeric effect that CH_2OH CH_zOH has higher level of stability than "=Br HO OH Br.

It has been noticed that anomeric effect is now taken into consideration as a special case of a general preference. It is considered as the generalized anomeric effect for the synclinal (gauche) conformations about the bond C-Y in the system X-C-Y-C. In the system X-C-Y-C, X, and Y are the heteroatoms that consist of the nonbonding electron pairs.

It has been noticed that usually at least one of the X and Y in the above-mentioned system is nitrogen, oxygen or fluorine. For instance, in the compound Chloro (methoxy) methane, the stability against the synclinal conformation is achieved by the anomeric effect. It is for the same reason that Me-P-I is preferred over H C.

However, in the alkyl glucopyranosides, the anomeric effect functions at two different sites:

i. along the endocyclic C–1 oxygen bond (endo-anomeric effect); and

ii. along the exocyclic C–1 oxygen bond (exo-anomeric effect).

CH_2OH CH_zOH HOH_q is preferred over the HoHq Me-Me. It has been observed that the opposite preference is usually asked for in some of the systems. For example, glycopyranosyltrialkyl ammonium salts are the systems that claim opposite preference. This opposite preference, that is claimed by some of the systems, is referred to as the reverse anomeric effect.

Anomers Diastereomers of glycosides, hemiacetals or other connected cyclic forms of sugars, or related molecules that vary in configuration only at C–1 of an aldose, C–2 of a 2-ketose, etc. This was earlier used to explain the stereochemistry of oximes and other connected systems. Antipodes is an obsolete synonym term that is used for enantiomers. In the modern chemistry, the usage of this term is strongly discouraged.

There are several trigonal bipyramidal structures in the stereochemistry. These trigonal bipyramidal structures are denoted by Apical, Basal, and Equatorial terms. One of the examples of these kind of structures is a five-coordinate trigonal bipyramid with phosphorus as central atom.

The term apical is used to denote to the two positions that are on the same line with the central atom or to the bonds that connect these positions with the central atom.

The term equatorial is used to denote the three equivalent bonds (or positions) that are present in a single plane which passes through the central atom. This plane is found to be at the right angles with the direction of the apical bonds. Here, it can be understood that the term 'apical' can also be used to denote the bonds pointing from the atom at or near the center of the base to the apex of a pyramidal structure.

The term 'Basal' is used to denote the positions that are present at or near the base of the pyramid. Also, the term 'Basal' is also used to refer to the bonds that connect positions at the base of the pyramid to the central atom of the base.

Also, there have been some instances where the apical bonds have been denoted by the 'term' axial (1994). The term 'axial' is used to describe addition or elimination reactions. There have been instances where the term axial has been used inappropriately and incorrectly to describe the absence of a rotation-reflection axis (alternating axis) in a molecule. In other words, the term 'axial' is incorrectly used to explain the meaning of chirality. Still, there have been instances in the traditional terminology, where the term 'axial' is used to denotes bonds and positions for asymmetric carbon atom, asymmetric synthesis, asymmetric induction, etc.

7.10.1. Asymmetric Carbon Atom

The traditional name given to a carbon atom which is connected to four different entities (atoms or groups) is called an asymmetric carbon atom. This term was coined by van't Hoff. For example, Cabd denotes Carbon attached to four different groups. So, 'C' in this compound is an asymmetric carbon. Asymmetric Induction is the term, in the traditional usage, used to describe the preferential formation in a chemical reaction of one enantiomer or diastereoisomer over the other as a consequence of the impact of a chiral feature present in the substrate, reagent, catalyst or environment of the reaction (Figure 7.6).

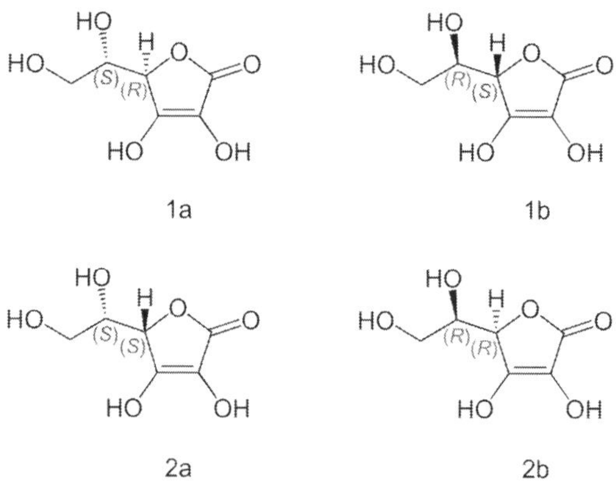

Figure 7.6. Various isomers of ascorbic acid giving an example of asymmetric carbon atom.

Source: Image by Wikimedia commons.

There are several processes by which the compounds containing asymmetric carbon can be produced. Asymmetric synthesis is a traditional term that is used for denoting these kinds of stereoselective synthesis of chiral compounds. After the production of the compound with asymmetric carbon compounds, it is important to purify the racemic mixture. Thus, the transformation or the conversion of a racemate is also known as the Asymmetric transformation. In order to purify the racemate, there are two possible pure forms, one form is the pure enantiomer or a mixture in which one of the enantiomers is present in excess. The other form of a pure mixture is in which the diastereoisomer is present in pure form or a single diastereoisomer is dominantly present in higher quantities as compared to the other diastereoisomer.

Some of the scientists and researchers denote this process by the term deracemization. If the two enantiomers of a chiral substrate A can be freely converted to other form and if an equal amount or excess of a non-racemizing second enantiomerically pure chemical species, say (R)-B, is added to a solution of racemic mixture A, then the final equilibrium mixture of adducts A*B will consist of unequal amounts of the diastereomers. This is a very general preposition for these kinds of mixtures.

The final result of this equilibration is called the asymmetric transformation of the first kind. In this kind of system, if the two diastereomeric adducts have some considerable variations in the solubility so that only one of them, say (R)-A*(R)-B, crystallizes from the solution, then the equilibration of diastereomers in solution and concurrent crystallization will continue so that all (or most) of the substrate A can be isolated as the crystalline diastereoisomer (R)-A*(R)-B.

This kind of an "asymmetric transformation induced by crystallization" is known as the asymmetric transformation of the second kind. A tropisomers A subclass of conformers which can be isolated as completely different chemical species and which is formed from the restricted rotation about a single bond e.g., ortho-substituted biphenyl, 1,1,2,2-tetra-tertbutylethane.

7.10.2. Many Simple Reactions Involve Stereochemistry

If only achiral reactants and reagents are involved in the reaction then the reaction can produce only racemic mixtures, or it can produce the products that are achiral. Therefore, a mixture of 1-bromobutane and 2-bromobutane is produced by the monobromination of butane. There is no stereogenic center in 1-bromobutane and thus it is achiral in nature.

There is a Stereogenic center in the 2-bromobutan. It is a mixture of the enantiomers (R)-2-bromobutane and (S)-2-bromobutane. This racemic mixture is produced because the reaction proceeds through a free-radical mechanism. It has to be noted that free radicals are almost perfectly planar. The product mixture is made achiral by this planarity. This is because there is equal possibility that a bromine free radical is attacking from any of the side of the carbon free radical intermediate.

The atom of halogen is added to an alkene proceeds through a trans addition. This takes place because of the steric hindrance that is made by the creation of the halonium ion.

It has to be noted that the catalytic hydrogenation of alkenes as well as alkynes occurs through cis addition. This is because the absorption of the atoms of the hydrogen on the surface of the rigid catalyst permits the atoms of the hydrogen atoms to approach the double bond from just one side of the alkene molecule.

The Dehydrohalogenation of chloroethane to form the ethene is a very important reaction. This reaction takes place through a trans elimination reaction. This is because of the tetrahedral shape of the atom of the carbon that is bonded to the chlorine atom. The electron pair of the other carbon must try to reach the chlorine atom from the unhindered side in order to displace the chlorine as a chloride ion.

7.11. STEREOCHEMISTRY OF NUCLEOPHILIC SUBSTITUTION REACTIONS AT SATURATED CARBON

Nucleophilic substitution at soaked carbon is a typical and helpful response type. These responses are found to happen by means of two particular unthinking sorts, which are assigned SN1 and SN2. The stereochemistry related to both of these robotic sorts will be talked about, starting with the SN2 response.

7.11.1. The Stereochemistry of SN2 Reactions

The SN2 reactions alludes to "substitution, nucleophilic, bimolecular." The assignment bimolecular alludes to the number of particles in the rate-deciding advance. Summed up instances of SN2 responses are spoken to beneath, perceiving that the nucleophile might be contrarily charged or uncharged and the leaving gathering (when present in the substrate) might be uncharged or emphatically charged.

Whatever the charge sort of the nucleophile or the leaving gathering, these responses perpetually happen with a complete reversal of design at the soaked (sp3) carbon experiencing substitution. One of the soonest (and particularly exquisite) showings of reversal of arrangement in an SN2 response is delineated in the response of optically unadulterated 2-iodooctane with radioactive iodide particle.

The watched reaction, the fuse of radioactive iodine into the substrate, is thoroughly second request (first request in iodide particle and first request in iodooctane), as it ought to be for an SN2 instrument. The key perception is that the pace of joining of radioactive iodine into 2-iodooctane is actually one-portion of the pace of racemization of the optically dynamic substrate. This must be the situation if every response of radioactive iodide particle with 2-iodooctane racemizes two atoms of this substrate.

Thus, this must be the situation if every response happens with reversal of design, the altered item, joined with a particle of the first item, at that point gives two atoms of racemic (R + S) 2-iodooctane. In the event that the response had happened by means of a carbocation course, racemization would have been the consequence of every response with radioactive iodide particle, making the pace of racemization precisely equivalent to the pace of radioactive iodide particle consolidation.

Obviously, the assurance of total setups of reactants and items presently makes the assurance of response stereochemistry moderately clear. The hypothetical inquiry of particular intrigue is the reason the inclination for reversal over speculative maintenance is so ground-breaking in these responses.

By chance, it obviously has nothing on a very basic level to do with the condition that an adversely charged nucleophile may electrostatically rebuff a leaving gathering, which in the atom is uncharged however which is added during the time spent getting contrarily charged in the progress state, in a retentive system, since the reversal is still unequivocally favored in any event, when the leaving gathering (when present in the particle) is decidedly charged and the entering nucleophile is contrarily charged (bringing about an electrostatic fascination).

The hypothetical basis for the exceptionally evolved inclination for an invertive response instrument can be seen by applying either the idea of change state aromaticity/antiaromaticity or by applying boondocks orbital (HOMO/LUMO) hypothesis.

The previous methodology has just been created, and for the particular instance of nucleophilic substitution. The clarification was given there was that the progress state for retentive substitution is unavoidably a cyclic one and it has four electrons (two from the nucleophile and two from the C-L bond), so it is antiaromatic, while the changing state for the invertive procedure is non-cyclic, with the goal that this TS is non-sweet-smelling.

How about presently thinking about the use of HOMO/LUMO (or FO) hypothesis, to the particular instance of bimolecular nucleophilic substitution. The electron pair being given plainly should originate from the nucleophile; so, the HOMO is the orbital of the nucleophile (Nu) which contains the unshared electron pair used to cling to carbon. In any case, shouldn't something be said about the LUMO?

Unmistakably, the substrate atom must stock the LUMO. Since the bond being broken is the C-L bond, the suitable LUMO must be that of this bond. Reviewing that this anti-bonding LUMO of the C-L bond has a hub in the inside piece of the bond, it is visible in the representation that though the nucleophile can cling to the carbon end of the bond, as it needs to, it can't simultaneously cling to the L part of the bond.

The spatially unavoidable communication with L is in truth anti-bonding, and will, in general, reduce the net holding connection of the nucleophile with the natural substrate. Conversely, if the nucleophile enters from the posterior of the C-L bond, it is very expelled from L and there is no huge anti-bonding connection to meddle with the holding communication between the nucleophile and carbon.

It is critical to understand that, in spite of the fact that holding collaborations between the nucleophile orbital and both the carbon and L orbital's are conceivable on the off chance that can be utilized for the C-L BMO, this connection would be between two filled orbital's, and along these lines can yield no net adjustment. Recall this is on the grounds that in the co-operations between two filled orbital's, there are four electrons to oblige.

7.12. IMPORTANCE OF STEREOCHEMISTRY IN CHEMICAL REACTIONS–THALIDOMIDE DISASTER

- It has to be noted that the arrangement of the atoms in three-dimensional space plays a very important part in the properties of the molecule;
- The example of the importance or the significance of stereochemistry can be observed in the thalidomide disaster. In the year 1957, it struck Germany;

- The drug thalidomide was sold as an over-the-counter drug. The initial intended use of this drug was to combat nausea. Pregnant women also make use of this drug to alleviate morning sickness;
- Though, it has been discovered that the drug has gone through racemization and it makes a mixture of enantiomers in the human body because of the process of metabolism;
- Among all the enantiomers, one of the enantiomers is believed to be the cause of genetic damage in developing embryos and it also leads to birth defects in infants;
- This is grounded on the data that more than 5000 babies all over the world were born which were having deformed limbs shortly after thalidomide was sold commercially as an over-the-counter drug;
- This unexpected impact of the drug led to the imposition of stricter laws for the drug regulation (it has been observed that just 40% of the babies who were born with these deformities survived); and
- The disaster shows why is it important to have a proper understanding of the stereochemistry.

7.13. CONCLUSION

Stereochemistry is the science of understanding the spatial arrangements. This branch of science helps in getting an understanding of the phenomena taking place in the chemical reactions. While understanding the spatial arrangements of the atoms in a chemical reaction is very important, it is equally important to understand how stereochemistry defines various important aspects in the chemical reaction. The retention and inversion of the stereochemical configuration is one of the most important aspect of the chemistry. The absolute configuration of the compounds with all the major aspects of angle strain, asymmetric carbon atom and various other simple reactions involving the stereochemistry explain the impact of different types of spatial arrangements of the atoms on a chemical reaction.

REFERENCES

Achmatowicz, O., Favre, H., Gile, P., Moss, G., Riche, Tavernier, D., Weissbach, O., et al., (1996). *Basic Terminology of Stereochemistry* (12th edn.). [ebook] Great Britain: G. P. MOSS. Available at: https://www.uni-bi.de/(de)/chemie/oc3sewald/downloads/200910_ws/IUPACrecommendations.pdf (accessed on 18 August 2020).

BYJUS, (n.d.). *Stereochemistry – Chirality, Enantiomers and Diastereomers with Videos.* [online] Available at: https://byjus.com/chemistry/stereochemistry/ (accessed on 18 August 2020).

Cliffsnotes.com. (n.d.). *Stereochemistry of Reactions.* [online] Available at: https://www.cliffsnotes.com/study-guides/chemistry/organic-chemistry-i/stereochemistry/stereochemistry-of-reactions (accessed on 18 August 2020).

Courses.lumenlearning.com. (2020). *4.6. Separating Enantiomers | Organic Chemistry 1: An Open Textbook.* [online] Available at: https://courses.lumenlearning.com/suny-potsdam-organicchemistry/chapter/4–6-separating-enantiomers/ (accessed on 18 August 2020).Courses.lumenlearning.com. (n.d.). *4.5. Stereochemistry of Reactions | Organic Chemistry 1: An Open Textbook.* [online] Available at: https://courses.lumenlearning.com/suny-potsdam-organicchemistry/chapter/4–5-stereochemistry-of-reactions/ (accessed on 18 August 2020).

Kennepohl, D., (2015). *5.7: Stereochemistry in Chemical Reactions - Chemistry Libretexts.* [online] Chem.libretexts.org. Available at: https://chem.libretexts.org/Bookshelves/Organic_Chemistry/Map%3A_Organic_Chemistry_(Vollhardt_and_Schore)/05._Stereoisomers/5.7%3A_Stereochemistry_in_Chemical__Reactions (accessed on 18 August 2020).

Research.cm.utexas.edu. (2020). *CH 391: Unit 3: Stereochemistry.* [online] Available at: http://research.cm.utexas.edu/nbauld/368_unit3.htm (accessed on 18 August 2020).

Toppr-guides, (n.d.). *Stereochemistry and Chemical Reactions of Haloalkanes: Videos, Q&A.* [online] Available at: https://www.toppr.com/guides/chemistry/haloalkanes-and-haloarenes/chemical-reactions-and-stereochemistry/ (accessed on 18 August 2020).

Zoomtute, (2019). *Stereochemistry is an Important Aspect of Organic Chemistry.* [online] Available at: https://www.zoomtute.com/content/blogs/13/stereochemistry-is-an-important-aspect-of-organic-chemistry (accessed on 18 August 2020).

Chapter 8

Topicity and Prostereo-Isomerism

CONTENTS

Topicity and prostereoisomerism (PIM) are the two important aspects of stereochemistry. In this chapter, both these factors are evaluated and their importance is described considering different examples. This chapter also evaluates how the previous approach related to prostereoisomerism were considered. Ogston theory is the most essential theory given in relation to the stereochemistry.

The chirality aspect for any molecule is related to the fact that a molecule or compound is chiral only when it is not able to superimpose with their mirror image. These aspects of the stereochemistry have been described in this chapter. The topicity is another aspect that has been described here which is the relationship among the substitutes and the structure to which they are attached. Four different relationship types of the topicity are also described here.

8.1. INTRODUCTION

Objects which are mostly molecules that are unable to superpose with the mirror image of the molecule, for this phenomenon the word used is Chirality. Molecules that are able to display the effect of chirality that is they do not superimpose with their mirror image are known as chiral, for example, (S)-(+)-lactic acid.

As per the chapter pro-stereoisomerism and prochirality are the same and thus, chirality, and prochirality have been evaluated for different compounds. There is a stereochemical relation among the substituents and the structure to which they are attached, this relationship is called topicity.

The major impact of the chirality of a molecule is the chiral center. In lactic acid, the carbon atom is the chiral center but for chirality, there are other elements too that are responsible such as the chiral axis as in allenes or the chiral plane, as in certain substituted para cyclophanes.

Often, one is keen on the fact that in specific particles, for example, propionic acid an achiral focus, CO can be changed into a chiral focus by substitution of one or other of two evidently indistinguishable ligands by an alternate one.

In this manner, the substitution of H_a at C_a in propionic acid by OH produces the chiral focus of (S)-lactic acid while the undifferentiated from the substitution of Hb offers to ascend to the enantiomeric (R)-lactic acid.

C_a in propionic acid is in this manner called a "prochiral focus" H_a and H_b are designated "heterotopic ligands." Prochiral axes and planes may also be characterized according to chiral planes and axes.

Substitution is one of two regular methods for interconverting organic elements, the other is expansion. The chiral focus in lactic acid can, on a basic level, be produced by the expansion of methyl magnesium iodide to the carbonyl gathering of glyoxylic acid.

Contingent upon the destiny of the aldehyde the Grignard reagent adds to, either (S)-or (R)-lactic acid is acquired. Subsequently, the carbonyl in glyoxylic acid is likewise said to be prochiral and to introduce two heterotopic faces.

The term prochirality, however, is a frequently used term among biochemists, but still, there is a limitation attached to it which is due to the fact that there is a limitation attached to the definition of chirality.

Atoms may show absolutely stereochemical contrasts without being chiral: cis-trans isomers of olefins and certain achiral cis-trans isomers of cyclones are models. Along these lines (Z)-and (E)–1,2-dichloroethylene (4, 5) are achiral diastereomers, as are cis-and trans-l,3-dibromocyclobutanes being without chirality these mixes have no chiral focuses (or other chiral components).

In this manner it is wrong to connect stereoisomerism with the event of chiral components; rather, it can be stated that it is showed by the presence of stereogenic components or components of stereoisomerism.

The idea of prochirality must be summed up to one of prostereoisomerism (PIM). It is exemplified by chloroethylene and bromocyclobutane, these particles show PIM in as much as substitution of the homomorphic molecules by chlorine offers to ascend to the stereoisomers individually.

Essentially, substitution of homomorphic molecules, separately, by bromine offers ascend to the stereoisomers. Therefore, it has a focal point of PIM (or prostereogenic focus) at C_3 and has a prostereogenic pivot concurring with the pivot of the double bond.

8.2. PREVIOUS APPROACH

From the substance perspective, the most noteworthy part of the current subject has for the possibility of separating the heterotopic ligands of components attached to PIM or the heterotopic appearances of a prostereogenic double bond.

The idea of heterotopic ligands and their acknowledgment, in reasonable cases, by NMR spec-troscopy was first introduced in a renowned article by Mislow and Raban, on which a great part of this work is based.

Separation of heterotopic ligands or appearances might be compound or biochemical (as in topsy-turvy and stereoselective blend, including changes by proteins) or spectroscopic (prominently by NMR spectroscopy).

Prior to entering upon these points in detail, it should be presented here a test to delineate the utility of the ideas: In the citrus extract, can the four methylene hydrogens H_a, H_b, H_c, H_d be recognized by NMR spectroscopy, or by ethicalness of their association in the enzymatic lack of hydration of citrus extract to cis-aconitic acid, or both?

This inquiry can, without much of a stretch, be addressed once the fundamentals of PIM are seen: all the hydrogens can be recognized by fitting enzymatic responses and H_a and H_b (just as H_c and H_d) can offer ascent to distinctive signals in the proton NMR range though H_a and H_c (or H_a and H_d) offer ascent to correspondent signs.

However, the principal significance, including prochirality (however not perceived accordingly) was the decarboxylation of methyl ethyl malonic acid to methyl butyric acid within the sight of brucine. The C_2 carbon in a-methyl butyric acid is a chiral focus; the C_2 carbon in the malonic acid antecedent is a prochiral focus.

The item is optically dynamic; for sure, this is one of the principals recorded unbalanced amalgamations. Plainly the shallow impression that the two carboxyl groups of the beginning malonic acid are proportionate must be incorrect, for they can, on a fundamental level, be recognized within the sight of the chiral catalyst brucine.

A superior archived case is that of citrus extract. It was for quite some time realized that when oxaloacetic acid is taken through the Krebs cycle, the c-ketoglutaric acid framed is named solely at C_1 (alongside the keto gathering), and not under any condition at C_5.

This discovering appeared to toss question until then expected intermediacy of citrus extract in the cycle since, it was contended, the two parts of the bargains (CH_2CO_2H) are comparable and in this way, the ketoglutaric acid shaped through this middle of the road ought to be named similarly at C_1 and C_5.

However, it is currently evident that the trial, not the slightest bit wipes out citrus extract as a potential halfway in the oxaloacetic acid ketoglutaric acid change by the goodness of the way that the two CH_2CO_2H branches are, indeed, unique, and recognizable by proteins as a result of the prochiral idea of the focal carbon molecule, C_3.

Additionally, the way that phosphorylation of glycerol with ATP within the sight of the chemical glyeerokinase gives solely (R) glycerol-l-phosphate shows that the compound can recognize the two essential alcohol groups of glycerol and that these gatherings should subsequently be particular: C_2 in glycerol is prochiral.

8.3. PROSTEREOISOMERISM

The idea of PIM (consequently 'PIM') is notable to organic stereochemistry students with regards to the memorable Ogston theory. In the broadest sense, PIM shows itself when certain achiral (visitor) particles are put in a chiral domain.

In the prototype case, the visitor atom has two indistinguishable and two nonidentical bunches attached to a tetrahedral focus. In an achiral situation, the two indistinguishable gatherings are vibrantly comparable, yet this decline is lifted in a chiral domain.

In this way, the cooperation with the chiral condition renders the proportionate gatherings nonequivalent, and the visitor particle, in essence, becomes chiral. This potential for procuring chirality in certain achiral atoms is caught by the term PIM.

The tetrahedral focus was therefore named 'prochiral,' in spite of the fact that the broader term 'prostereogenic' is as of now liked. PIM may likewise be shown by unsymmetrical planar atoms, the two planes (faces) of which become prostereogenic in a chiral domain.

Besides, PIM is frequently defined as far as a demonstration of segregation between equal gatherings or appearances, normally by means of a response prompting stereoisomeric items. In any case, PIM can likewise be observed by a basic demonstration of review conceivable in light of the fact that the human performing it has a chiral sense.

8.3.1. Considerations Related to Symmetry

PIC particles clearly define a center ground among 'achiral' and 'chiral.' As found in various cases with a tetrahedral focus, the nearness of two nonidentical groups is basic; these particles despite everything have a plane of symmetry, in spite of the fact that defining a subclass among achiral atoms.

The quality of the connection between the visitor atom and the chiral condition would, obviously, rely upon different variables, yet especially on the idea of the groups and the chiral condition, the number of communications, and so forth.

The accompanying 'Gedanken tests' offer fascinating bits of knowledge into the sign of PIM under different conditions. Consequently, even a profoundly balanced atom, for example, methane (CH_4) may turn into a chiral moiety inside the achiral condition.

This necessitates the particle to be basically limited from tumbling, which would 'scramble' the overall places of all the four hydrogen molecules. However, if the particle is obliged inside a chiral pocket, all the four hydrogen molecules become not equal: they would at that point, on a fundamental level, show four unique resonances in the proton NMR range.

Putative connection of the CH_4 molecule at one of its hydrogens to the chiral pocket would likewise prompt the nonequivalence of the staying three hydrogens, as long as the pace of pivot of the CH_3 were delayed on the NMR time scale (Figure 8.1).

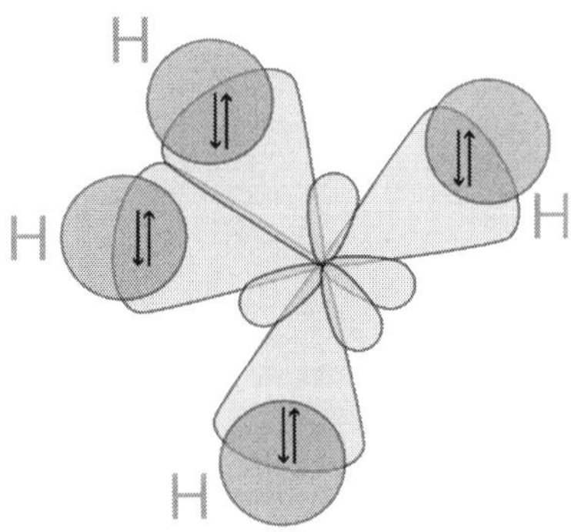

Figure 8.1. Hybridization of CH_4 molecule.

Source: Image by Wikimedia.

It is an interesting fact that this stands out from the instance of a PIC atom, basically in view of its lower symmetry. In this manner, the identical gatherings in a PIC particle would become non-proportionate in a chiral situation regardless of whether the included atom was uninhibitedly tumbling.

This is because of the nearness of the two non-identical gatherings, which give a 'stereochemical fix' at the tetrahedral focus. In this way, any tumbling occasion that exchanges the places of the comparable gatherings inside the chiral condition additionally trades the places of the non-proportional gatherings.

Since the last trade creates putative diastereomeric collaborations (with the chiral condition), tumbling can never prompt the dissolution of the particle.

8.3.2. Ogston Hypothesis and Two-Dimensional Chirality

The collaboration of an achiral atom with a chiral domain prompts results that rely basically upon the balance level of the particle, yet additionally on the idea of the cooperation (the atom being compelled).

In any case, apparently, the idea of the chiral condition is significant, a wise end that was evidently forecasted by the Ogston theory, regularly thought to be a milestone in the historical backdrop of stereochemistry.

Interestingly, in spite of the fact that the proposal led to the development of the possibility of PIM itself, a couple of its nuances evidently stay to be expounded. It currently creates the impression that the possibility of two-dimensional chirality is critical to understanding the Ogston theory, vide infra.

The first Ogston proposition goes back to a period when current stereochemical thoughts were all the while developing, and the idea of a chiral domain inside a pocket of the host atom would maybe have been a curiosity.

The speculation is best exemplified by the enzymatic phosphorylation of glycerol to create R-(–)-glycerol-1-phosphate, which includes recognizing the two terminal hydroxyl bunches in the PIC glycerol particle.

In spite of the fact that the development of a chiral item at a chiral compound dynamic site needs small anguishing, it is a demonstration of the savvy of the methodology that robotic subtleties were for sure sought after.

Furthermore, this is likewise a fortunate intersection refer to a remarkable early paper by Easson and Stedman, which excessively forecasted a large number of the later improvements identified with the Ogston theory and PIM, generally.

These creators expected a three-guide restricting mode all together toward relating the organic action of different particles as far as their stereochemistry.

8.3.3. Chiral Surfaces and Planar Chirality

The mechanism of Ogston included a three-guide connection of a PIC atom toward, basically, a chiral surface, for example, a chiral pocket in which one face is protected and consequently distant. At that point, at least three cooperation is important unequivocally to exclude one of the two comparable groups at the prostereogenic focus from the response site.

Subsequently, a two-point connection would not recognize the two proportionate groups, in as much as the unattached groups would not cooperate with the chiral surface. Note likewise that it takes at least three focuses to define a surface.

Consequently, the Ogston theory summoned what is currently comprehended as two-dimensional chirality. This is conceivable on account of an unsymmetrical planar object as long as the item isn't moved out of the plane (Figure 8.2).

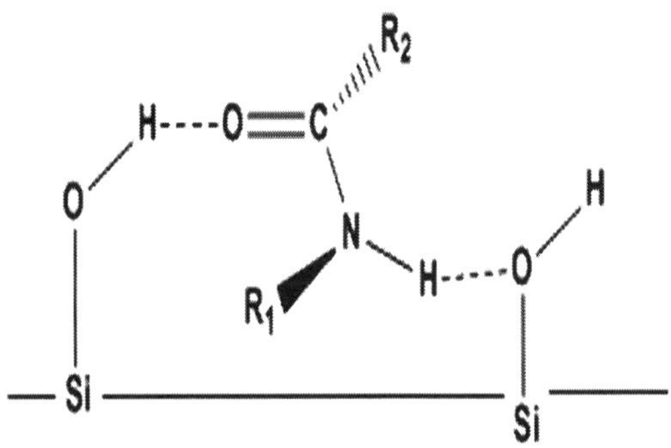

Figure 8.2. Chiral surface example amide on silica.

Source: Image by Wikimedia.

Hence, inside the general plane involved by it, the unsymmetrical planar item would be nonsuperimposable on its perfect representation. This 'stereochemical chiral surface, however, secures accessible significance when a chiral pocket works as an accepted surface.

Then again, the Ogston speculation suggests that coupling destinations in organic responses might be seen basically as chiral surfaces. However, in the light of current information about the three-dimensional structures of organic macromolecules, the Ogston theory adds up to 'pointless excess.'

In this manner, a thought of the general three-dimensional space inside a chiral (have) pocket demonstrates that the two comparable gatherings at a prostereogenic focus in a visitor atom might be recognized regardless of whether the particle was to be unreservedly tumbling.

Obviously, restricting connections would (for the most part) increment the vitality difference between the elective bound gatherings at the response community. On a fundamental level, in any case, they are a bit much if the spatial condition in the chiral pocket is sufficiently requesting as far as selectivity.

Then again, such a perspective on stereoselectivity should be tempered by ordinary contemplations including current information and experience. In this manner, have visitor communications by and large include exact restricting locales prompting the arrangement of the middle of the road buildings of defined structure.

These may undoubtedly speak to minima along response shapes in potential vitality space. On this premise, we are currently back at the Ogston theory.

In this way, in spite of the fact that the immobilization of a PIC substrate at a catalyst dynamic site is, on a fundamental level, pointless so as to recognize the two comparable gatherings (or countenances), the truth of the matter is that responses at such destinations do happen through bound intermediates.

After considering all things, the inquiry emerges with regards to the base number of connections that are important to recognize the equal groups. The appropriate response would be 'three,' expecting that the coupling site can be considered as a chiral surface.

The perspective on a compound dynamic site as a chiral surface seems unexceptional, taking into account that the coupling site to some degree looks like a mousetrap, for example, one side is totally protected, so the section and exit of the substrate are limited to the contrary side.

This may appear to be an over simplification, because of the fact that it disregards powerless dispersive associations (e.g., van der Waals) between the substrate and the dividers of the dynamic site parted.

In any case, the Ogston instrument plainly centers around the three most grounded connections as three useful destinations in the substrate are included. The above mousetrap similarity needs due alert, as demonstrated by a captivating paper by Mesecar and Koshl and on isocitrate dehydrogenase.

Right now, a dynamic site is a three-dimensional split that is reciprocal to a speculative trigonal bipyramid. This can oblige a similar three (of the four) bunches connected to the C_2 stereogenic focus in the substrate enantiomers, however from inverse sides of the trigonal surface.

Basically, the official of the fourth group at C_2 then decides stereospecificity, demonstrating a takeoff from the Ogston three-point component. Hence, the compound can tie either L-isocitrate or D-isocitrate, in spite of the fact that the last alongside Mg_2+.

Interestingly, this may well show that three-point restricting applies to PIC substrates and four-direct officials toward chiral substrates. Obviously, a PIC substrate may prompt a chiral item (and the other way around), taking note of that proteins to tie the two reactants and items (whatever the heading of the balance).

In any case, the catalyst could then experience conformational changes during the response, and shift back and forth between the three-point and four-point modes as proper. Besides, truth be told, these particular restricting modes likely demonstrate that efficient restricting depends on size variety between the bound gatherings, as contended beneath.

Consequently, a chiral substrate has four different bunches at a tetrahedral focus, and a blend of any three of these may well offer a comparable level of size variety.

The three-point restricting mode would not then recognize the two enantiomeric substrates efficiently. Of four gatherings A, B, C, and D, both the mixes A, B, C, and B, C, D, state, could tie with comparative efficiency; regarding size, A <B<C<D.

8.3.4. Chiral Binding Sites

These contentions have fascinating results in the light of the way that chirality assumes a critical job in biomolecular collaborations. This is commonly decrypted with regards to selectivity and specificity, suggesting that, between two enantiomers of a substrate an organic restricting site especially interfaces with just one of them.

Nonetheless, these contemplations infer that a chiral restricting site likes to interface with a chiral instead of an achiral substrate particle. This depends on clear contemplations of shape complementarity (Figure 8.3).

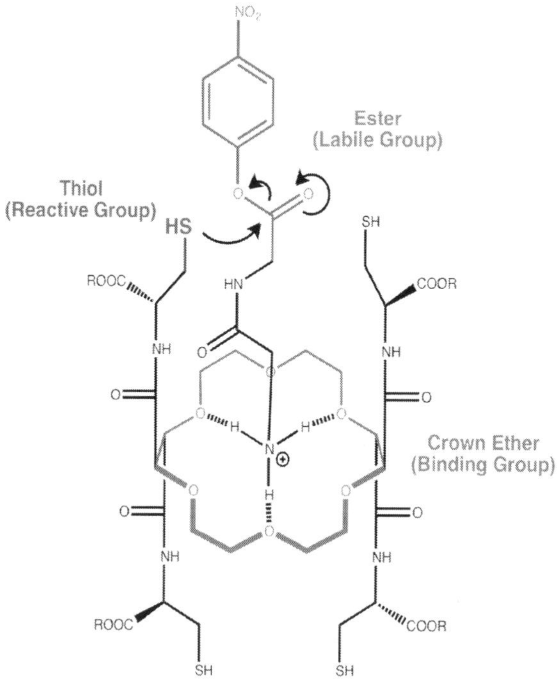

Figure 8.3. Chiral binding site: Chiral substituted crown ether catalyst.

Source: Image by Wikimedia.

In this way, a chiral restricting site might be defined as far as four restricting pockets appropriated in three-dimensional space. The pockets would be situated at the edges of a tetrahedron, for shape complementarity relative to a chiral tetrahedral focus.

Besides, the pockets would all differ, in either size or the idea of authoritative (chelation, hydrogen holding, and so on.). Let us assign the coupling pockets as far as (state) size as S (small), M (medium), L (huge), and XL (extra huge). Presently a tetrahedral substrate would perforce have four gatherings with similar qualities, and subsequently, be chiral.

8.3.5. Planar Systems

The thoughts might be considering the fact of planar atoms having prostereogenic faces, in spite of the fact that with extra warnings identified with their consistent qualities. These get from the prerequisite that these planar frameworks be reciprocal to a chiral surface, for example, they basically have a similar consistent quality (Figure 8.4).

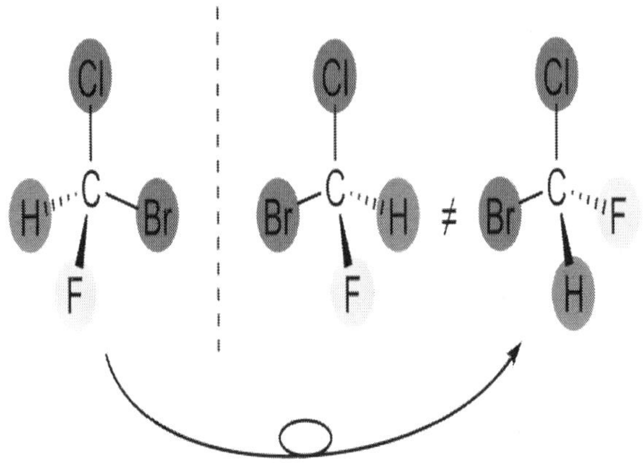

Figure 8.4. Chiral molecules example.

Source: Image by Wikimedia.

As a chiral surface, by definition, can't be separated from its plane ('flipped over'), this principle out the proximity of a C2 hub of consistency, e.g., naphthalene. Hence, these particles are basically described by unsymmetrical forms that offer ascent to edges of changing size and shape.

The particles may tie on a chiral surface that is defined by specialties along with the encompassing dividers of a chiral gap that are reciprocal to these edges. This would then render the top and base features of the atom nonequivalent or heterotopic.

As noted above, PIM can be considered as accepted chirality concerning an outside chiral surface. On account of tetrahedral frameworks, the three nonequivalent gatherings around the prostereogenic focus can be spoken to as S, M, and L.

On account of planar frameworks as well, the edges along the unsymmetrical shape can be partitioned into areas of shifting size dependent on flow. The previously mentioned standard again demonstrates that a PIC particle would be a suitable substrate at a chiral surface.

8.4. HOMOTOPIC LIGANDS AND FACES

It has been indicated that some apparently alike ligands are, in fact, not equivalent to enzymes or in their NMR signals. How does one decide, then, whether nuclei are equivalent? There are two criteria, a substitution criterion and an asymmetry criterion. Similar criteria (addition or symmetry) serve to test the equivalency of faces.

8.4.1. Substitution and Addition Criteria

Two homomorphic ligands are homotopic in the event that substitution of the initial one and the other by an alternate ligand prompts a similar structure. In this manner, the two hydrogen particles in methylene chloride are homotopic in light of the fact that substitution of either by bromine gives the equivalent $CHBrCl_2$ atom.

The three methyl hydrogens in acidic corrosive are homotopic on the grounds that substitution of any of them by chlorine gives very much the same chloroacetic corrosive; the two methine hydrogens in (R)-(+)-tartaric corrosive are homotopic on the grounds that substitution of both of them, for example, by deuterium gives the equivalent (2R, 3R)-tartaric-2-d corrosive.

Two comparing features of an atom (for the most part yet not continuous faces of a double bond) are homotopic when the expansion of a similar reagent to either confront gives a similar item.

For instance, expansion of HCN to acetone will give a similar cyanohydrin, regardless of which face expansion happens and expansion of bromine to ethylene correspondingly gives $BrCH_2CH_2Br$, paying little attention to the essence of the approach.

The two essences of the C = O double bond of acetone and of the C = C double bond of ethylene are hence considered homotopic.

8.4.2. Symmetry Criterion

Ligands are homotopic in the event that they can trade place through the activity of a C_n symmetry pivot. Hence the chlorine atoms in methylene chloride (symmetry point group C_{Ev}) are homotopic since they trade place through a 180° pivot the C_2 axis.

Similar to that, the methine hydrogens of (+)-tartaric corrosive are traded by the activity of the C_2 axis (the atom has a place with point group C_2). The circumstance in acidic corrosive is to some degree increasingly entangled. In the event that the atom is portrayed as stationary in one of its dominated adaptations.

However, turn around the HaC-CO_2H pivot is fast on the time size of most analyses. We are subsequently managing an instance of arrived at the midpoint of symmetry prompting trade of the three methyl hydrogens of CH_3CO_2H, which are in this manner homotopic when the revolution is quick on the time size of whatever test is being thought of.

Recognize that the proximity of a symmetry axis in an atom doesn't ensure that homomorphic ligands will be homotopic. It is essential that the activity of the symmetry pivot makes the cores being referred to trade places.

In this way in 1,3-dioxolane, in its normal planar adaptation, the hydrogens at C_2 are homotopic since they are exchanged by the activity of the C_2 axis (the symmetry point gathering of the particle is C_2.

Then again, the geminal hydrogen atoms at C_4 [or C_5] are not interconverted by the C_2 symmetry activity and are in this manner heterotopic. Anyway, H_a, and H_d are homotopic (as are H_b and H_c), being traded by the C_2 axis.

8.5. TOPICITY IN STEREOCHEMISTRY

The subfield of stereochemistry is the part of science that is related to the three-dimensional or spatial direction of compounds. Under the umbrella of stereochemistry, topicity tries to characterize the particular connection between particles or a group of atoms that are attached to an atom.

It's fundamentally the same as how one characterizes the family members in their family on the basis of their situation in the family tree.

8.6. TOPICITY RELATIONSHIPS

The topicity of two or more substituents is classified as one of four different relationships, which we'll be looking at next. The four possible relationships are called homotopic, heterotopic, enantiotropic, and diastereotopic.

8.6.1. Homotopic Relationships

In the event that two substituents inside a particle are proportionate (same), they are said to be homotopic. It is like two brothers in a family being indistinguishable twins. A genuine case of this relationship would be the hydrogen particles of methane. The hydrogens in methane are homotopic on the grounds that they are no different than each other, regardless of how a person controls the atom.

8.6.2. Heterotopic Relationships

Suppose that one had two sweet-smelling compounds and they contrasted in the manner that the hydrogens were bonded to the benzene ring. Since the two hydrogens named in red are at various situations on the ring, these two hydrogens are heterotopic to each other. These are instances of established isomers (same substance equation, diverse particle availability).

8.6.3. Enantiotropic Relationships

Presently, considering a compound in which two hydrogens (marked in red and blue) are attached to a carbon that contains a liquor group (-OH) and a sweet-smelling ring (Ph gathering). Suppose one supplanted the red hydrogen with a conventional atom, call it 'X', for example (Figure 8.5).

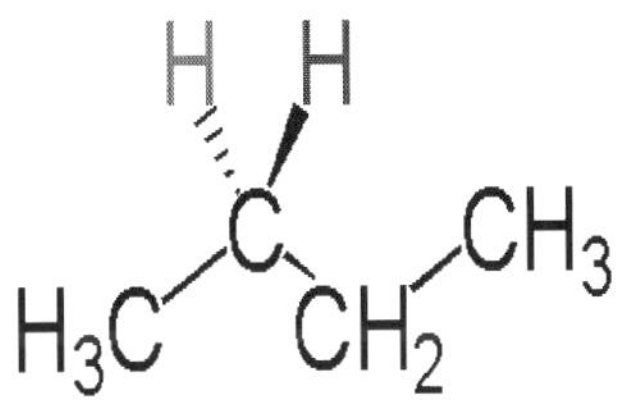

Figure 8.5. Enantiotropic relationship.

Source: Image by Wikipedia.

8.7. HETEROTOPIC LIGANDS AND FACES IN ENZYME-CATALYZED REACTIONS

8.7.1. Heterotopicity and Stereoelective Synthesis

Substitution of stereoheterotopic groups or expansion to stereoheterotopic faces offered ascend to stereoisomers. The rate of such substitutions of either of two ligands or augmentations to either of two appearances are frequently not the equivalent.

Specifically, substitutions of diastereotopic ligands or increments to diastereotopic faces, as a rule, continue at various rates in light of the fact that the progress states for such substitutions or augmentations are diastereomeric and in this manner are inconsistent in strength (Figure 8.6).

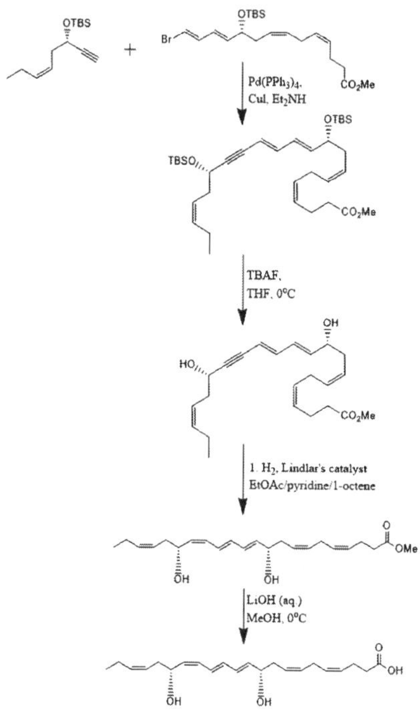

Figure 8.6. Convergent stereoselective synthesis.

Source: Image by Wikipedia.

Accordingly, the responses not just offer ascent to diastereomeric items, based on which ligand or face is included, however they give these items in inconsistent, in some cases very inconsistent sums, that is, they show diastereo selectivity (Izumi, 1971) or "diastereo separation" (Izumi and Tai, 1977).

Substitution of enantiotropic ligands or expansion to enantiotropic faces offers rise to enantiomeric items, however here substitution of the two ligands or expansion to the two faces commonly happens at a similar rate.

In light of the fact that the appropriate change states are enantiomeric and subsequently are equivalent in strength. This circumstance changes, nonetheless, when the reagent (or other substance taking an interest in the progress state, for example, the dissolvable) is chiral.

In that situation, the two change states will become diastereomeric and the two enantiomeric items will be framed at inconsistent rates and in inconsistent sums: the response will be enantio selective (Izumi, 1971) or "enantio differentiating" (Izumi and Tai, 1977).

Right now, prochiral beginning materials offer ascent to nonracemic chiral items, one talks about enantioselective (or, less properly, topsy-turvy) blends.

8.7.2. Topological Relation and Stereoselectivity

Topological relations between ligands on the tetrahedral C particle and between faces around the plane of the trigonal C molecule are of prime significance for getting stereoselectivity. The terms topological, point, and topicity begin from the Greek word topos, the spot.

It presents particles that have protons on the tetrahedral carbon with different topological qualities. It additionally presents particles that have a planar carbonyl group or trigonal carbon atom and different topological qualities of the features around the C=O double bond. The balance connection for the hydrogen atoms Ha and Hb on indicated methylenic groups is extraordinary.

8.7.3. Applications to Enzyme-catalyzed Reactions

When one discovers which of two stereoheterotopic ligands or features of a substrate is associated with a chemical catalyzed response, one is in a situation to offer an important expression regarding the area of the substrate compared to the dynamic site of the catalyst.

While contemplations of PIM are in this manner valuable in clarifying the catalyst substrate relationship in the actuated complex of a chemical intervened response, it should likewise be focused on that such contemplations in themselves are deficient to give the total picture and that they should essentially be enhanced by numerous different methods in compound science.

The writing in the region of PIM in compound responses is tremendous and we should restrict ourselves right now the conversation of a couple of delegate models.

The chapter is started with an old-style explore identified with the stereochemistry of oxidation of ethanol and a decrease of acetaldehyde interceded by the chemical yeast alcohol dehydrogenase within the sight of the oxidized (NAD) and diminished (NADH) structures, separately, of the coenzyme nicotinamide adenine dinucleotide (Figure 8.7).

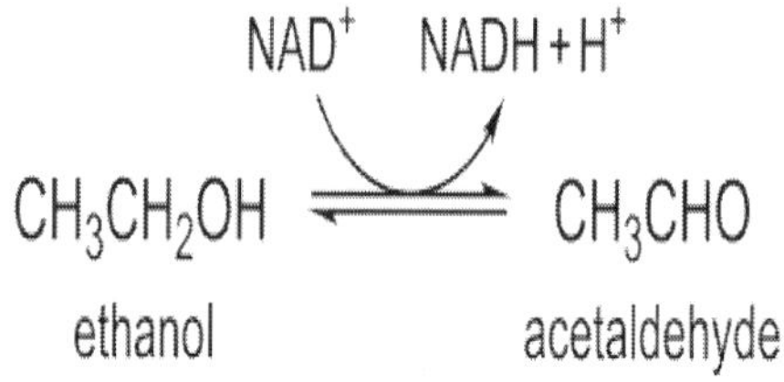

Figure 8.7. Ethanol to acetaldehyde.

Source: Image by Wikipedia.

The stereo chemically fascinating component of this response comes from the way that the methylene hydrogens in CH_3CH_2OH and the essences of the carbonyl in $CH_3CH{=}O$ are enantiotropic.

The inquiry hence emerges which of the CH_2-hydrogens is expelled in the oxidation and to which of the C=O faces the hydrogen appends itself in the decrease within the sight of the compound and coenzyme.

Westheimer, Vennesland, and Loewus in 1953 discovered that decrease of ethanol 1-d with NADH within the sight of yeast alcohol dehydrogenase gave ethanol-l-d which, upon enzymatic reoxidation by NAD +, returned ethanal-1-d without loss of deuterium.

There is subsequently a “stereochemical memory impact” associated with this response: The H and D of the CH_3CHDOH don’t get mixed however a similar H which is presented in the decrease is the one evacuated in the oxidation.

In the light of current knowledge, this is, obviously, not astonishing since the two methylene hydrogens are recognizable by bearing an enantiotropic relationship.

At the point when the setup of the ethanol-l-d is upset by change to the tosylate followed by treatment with hydroxide and the rearranged ethanol-l-d is then oxidized with yeast alcohol dehydrogenase and NAD +, the deuterium (which has taken the stereochemical position of the first hydrogen) is currently evacuated and the item is unlabeled $CH_3CH = O$.

The arrangement of occasions is abridged. Later analyses for a bigger scope built up that the ethanol-l-d got from $CH_3CD{=}O$ and NADH was levorotatory combined with the explanation of the arrangement of (--)-ethanol-l-d as S prompts the stereochemical picture condensed.

It follows along these lines that the hydrogen moved from the NADH in the enzymatic decrease appends itself to the Re face of the aldehyde and that this hydrogen consequently becomes HR in the ethanol; it is HR (the professional R carbinol hydrogen), thus, which is disconnected by NAD in the oxidative advance.

Plainly ethanol (and acetaldehyde) must fit into the dynamic site of yeast alcohol dehydrogenase so as to fit in with these stereochemical discoveries. A model for the decrease of a fundamentally the same as substrate, pyruvic corrosive (which is diminished by NADH within the sight of liver alcohol dehydrogenase to (S)-lactic corrosive) can be picturized.

Here we can recognize Ogston's image of the three-point contact, one contact being built up by the salt bond pyruvate-arginine-H +, the second by the hydrogen bond (histidine) N--H. O--C (pyruvate) and the third one including conveyance of the hydrogen of NADH (bound to the catalyst) to the Re face of the C = O of pyruvate.

The decrease of acetaldehyde is most likely comparable however the nonattendance of the COO group requires the contact at the third site to be built up in an alternate way.

It isn't sure that covalent or ionic holding is really associated with this contact; the state of the catalyst pit itself (and the specialist hydrophilic and hydrophobic collaborations between specific pieces of the compound and parts of the substrate) may add to the necessary direction of the substrate.

The investigation of the stereochemistry of ethanol oxidation and acetaldehyde decrease and the data identifying with the geography of the catalyst got from this examination are ordinary of an enormous number of different examinations of this type.

It has been examined here that the exchange of the hydrogen to and from the coenzyme includes its very own stereo chemical issue: In the reductive advance, is it HR or Hs of the dihydronicotin amide moiety which is moved from the coenzyme to the substrate; correspondingly, in the oxidation step, is the hydrogen disconnected from the substrate added to the Re or Si face.

Thermodynamic information on chemical catalyzed responses assume a basic job in the forecast of the degree of response and the situation of harmony for any procedure where these responses happen.

The significance of understanding the thermodynamics of these biochemical responses was underlined by Krebs and Kornberg 47 years prior in their monograph A Survey of the Energy Transformations in Living Matter (Krebs and Kornberg, 1957).

Their monograph contains a valuable reference section on Gibbs free vitality information of natural intrigue and a few information on the thermodynamics of enzyme catalyzed responses.

Key to both thermodynamics and to bioinformatics is the exact definition and determination of physical amounts, consistency in use, and institutionalization of names and images. Right now, TECRDB has clung to IUPAC proposals for biochemical thermodynamics (Alberty et al., 1994) and, generally, to the biochemical names prescribed by the Nomenclature Committee of the IUBMB (Webb, 1992).

Additionally, the requesting of the information in the TECRDB is done as per the Enzyme Commission number relegated to the compound used to catalyze the response considered. The Thermodynamics of Enzyme-catalyzed Reactions Database (TECRDB) is a far-reaching assortment of thermodynamic information on compound catalyzed responses.

The information, which comprise of obvious balance constants and calorimetrically decided molar enthalpies of response, are the essential exploratory outcomes collected from thermodynamic investigations of biochemical responses.

8.7.4. Retrosynthesis and Stereochemical Aspects of Synthetic Reactions

From an achiral particle, one enantiomer of the chiral target atom is shaped in an unbalanced response, otherwise called enantioselective response. Retrosynthetic investigation considers chiral target atoms as racemates and at this stage doesn't think about uneven combination to one enantiomer.

At the point when the second stereogenic focus in the chiral particle is presented in a stereoselective way, one of two conceivable diastereomers is shaped. Such an engineered response is called diastereoselective.

For simplification, during retrosynthetic investigation the arrangement of diastereomers isn't identified with the optical immaculateness of the objective particle. This implies that original chiral particles are considered racemates, which in a diastereoselective step manages one of the two potential racemic items.

Retrosynthetic analysis principally considers the electronic properties of synthons, i.e., the balancing impact of neighboring groups on the charged species in the ground state. This methodology isn't manageable to examination of unbalanced amalgamations since their result is controlled by the stereo electronic properties of the change state.

Stereoselective responses are actively controlled, and the discerning way to deal with examination of the stereoelectronic properties of the vigorously liked progress state is troublesome and typically supported by specific programs of computer.

Diverse stereoelectronic properties of comparative groups in responding particles bring about chemoselectivity, with higher reactivity of one group when contrasted with others.

This distinction is moderately simple to foresee over the span of retrosynthetic investigation. The result of an asymmetric response is substantially harder to conjecture considering the trouble in planning a progress state structure in transit to the favored enantiomer.

So as to recognize protected or basic isomers and stereoisomers it has to be seen that auxiliary isomers don't have a similar network. Stereoisomers, thus, have a similar network, yet they are connected or not as an item and its identical representation, i.e., they are enantiomers or diastereomers.

The arrangement is characterized as the general situation of iotas in the space that describe a solitary stereoisomer. The idea of a flat-out setup is especially significant in stereochemistry.

8.8. PROCHIRAL CARBONS

At the point when a tetrahedral carbon can be changed over to a chiral center by changing just one of the connected groups, it is alluded to as a 'prochiral' carbon. The two hydrogens on the prochiral carbon can be depicted as 'prochiral hydrogens.'

Note that if in a 'psychological study,' we were to change both of the prochiral hydrogens on a prochiral carbon center to a deuterium (the 2H isotope of hydrogen), the carbon would now have four diverse substituents and hence would be a chiral focus.

Prochirality is a significant idea in organic science since catalysts can recognize the two 'indistinguishable' groups bound to a prochiral carbon focus because of the way that they possess various areas in three-dimensional space (Figure 8.8).

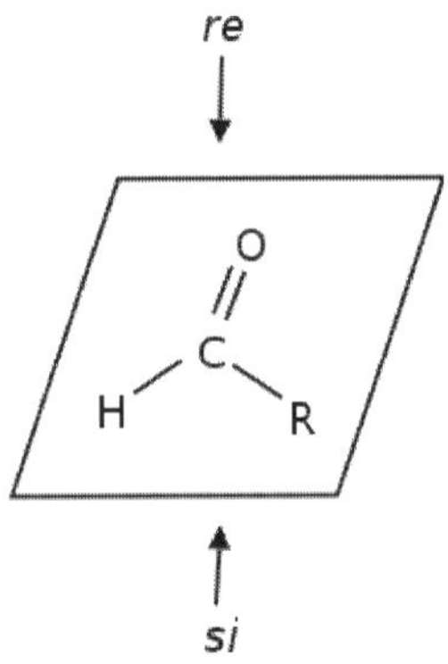

Figure 8.8. Example of prochiral carbon.

Source: Image by Wikimedia.

Consider the isomerization response beneath, which is a piece of the biosynthesis of isoprenoid mixes. The response did not have to be itself comprehended; the observation made now is that the isomerase chemical can recognize the prochiral 'red' and the 'blue' hydrogens on the isopentenyl diphosphate (IPP) substrate.

Throughout the left to right response, IPP explicitly loses the 'red' hydrogen and keeps the 'blue' one. Prochiral hydrogens can be unambiguously assigned utilizing a minor departure from the R's framework for naming chiral focuses.

For clearness, an exceptionally straightforward atom will be considered, which ethanol to clarify this framework is. To name the 'red' and 'blue' prochiral hydrogens on ethanol, the psychological study is required.

In the event that the requirement was to discretionarily change red H to a deuterium, the atom would now be chiral and the chiral carbon would have the R setup (D has a higher requirement than H). Hence, reference can be made to the red H as the ace R hydrogen of ethanol, and mark it HR.

On the other hand, if blue H is changed to D and leave red H as hydrogen, the configuration of the molecule would be S, so reference can be made to blue H as the pro-S hydrogen of ethanol, and it can be labeled as HS.

Glancing back at the isoprenoid biosynthesis model, it can be seen that it is explicitly the expert R hydrogen that the IPP substrate loses in the response. Prochiral hydrogens can be assigned either enantiotropic or diastereotopic.

If either HR or HS on ethanol was supplanted by a deuterium, the two coming about isomers would be enantiomers (in light of the fact that there are no different stereocenters anyplace on the particle). Accordingly, these two hydrogens are alluded to as enantiotropic.

In (R)-glyceraldehyde-3-phosphate ((R)-GAP), something else can be seen:

(R)-GAP as of now has one chiral focus. On the off chance that both of the prochiral hydrogens HR or HS is supplanted by a deuterium, a second chiral focus is made, and the two coming about atoms will be diastereomers (one is S, R, one is R, R).

At last, hydrogens that can be assigned neither enantiotropic nor diastereotopic are called homotopic. In the event that a homotopic hydrogen is supplanted by deuterium, a chiral center isn't made. The three hydrogen atoms on the methyl (CH_3) group of ethanol (and on any methyl group) are homotopic.

Groups other than hydrogens can be considered prochiral. The alcohol underneath has two prochiral methyl groups-the red one is star R; the blue is expert S. Citrate is another model. The carbon at the center is a prochiral focus with two 'arms' that are indistinguishable aside from that one can be assigned pro-R and the other pro-S.

In an isomerization response of the citrus extract (Krebs) cycle, a hydroxide is moved explicitly to the pro-R arm of citrate to shape isocitrate: once more, the chemical catalyzing the response recognizes the two prochiral arms of the substrate.

Albeit an alkene carbon clung to two indistinguishable groups isn't viewed as a prochiral focus, these two groups can be diastereotopic. H_a and H_b on the alkene underneath, for instance, are diastereotopic: if one of them is changed, and afterward the other, of these hydrogens to deuterium, the subsequent mixes are E and Z diastereomers.

8.9. PROCHIRAL CARBONYL AND IMINE GROUPS

Trigonal planar, sp^2-hybridized carbons are not, as we surely understand, chiral focuses yet they can be prochiral focuses if they are clung to three diverse substituents. It (and the proteins that catalyze responses for which they are substrates) can be recognized that the two planar 'faces' of a prochiral sp^2-hybridized groups.

These features are assigned by the terms r_e and s_i. To figure out which is there and which is the s_i face of a planar natural gathering, we essentially utilize a similar need ranking that is known from the R's framework and follows a circle: r_e is clockwise and s_i is counterclockwise.

8.10. DIASTEREOTOPIC LIGANDS AND FACES

The earlier mentioned criteria may also be employed to spot diastereotopic ligands, i.e., ligands which find themselves in a stereochemically distinct but non-mirror image environment.

8.10.1. Substitution-Addition Criterion

Since C_2 in 2-bromobutane is chiral, H_a, and H_b can't be enantiotropic and the substitution rule unveils that they are diastereotopic as opposed to homotopic. The instances of cyclobutanol and 4-t-butyl-l,l-difluorocyclohexane appear that a chiral focus isn't required for the presence of diastereotopic cores (Figure 8.9).

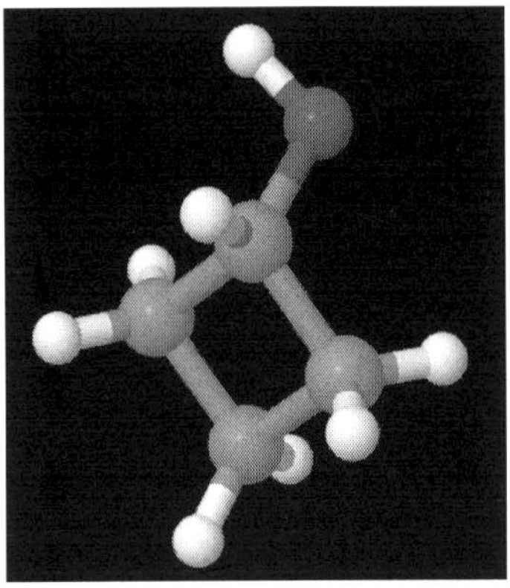

Figure 8.9. Cyclobutanol.

Source: Image by Wikimedia.

H_a and H_b in 48 and Fa and F_b are diastereotopic in light of the fact that they are cis- and trans, individually to the hydroxyl bunch at C_1 in or the t-butyl bunch at C_4. It may be noticed that, after the substitution, C_3 or C_1 isn't a chiral focus however a stereogenic focus; the relating particles are prostereogenic.

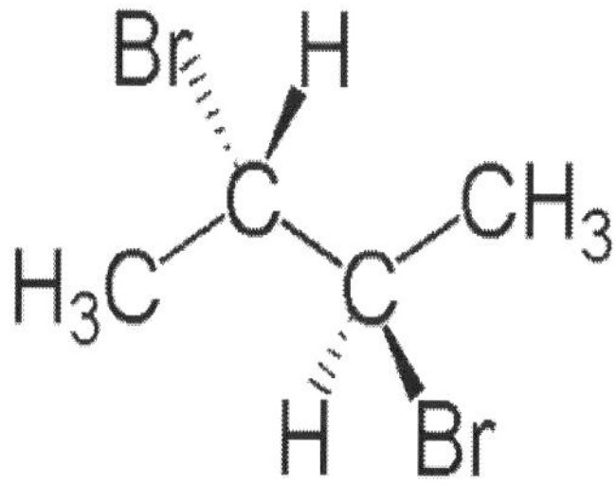

Figure 8.10. Diastereotopic ligand.

Source: Image by Wikipedia.

On account of propene substitution of H_a and H_b produces a cis-trans pair of (diastereomeric) olefins again making H_a and H_b diastereotopic. One is cis to the methyl bunch at the distal carbon particle, the different trans (Figure 8.10).

The instance of meso-2,4-pentanediol is of intrigue in light of the fact that the results of substitution of H_a and H_b are diastereomeric meso frames in which C_3 is pseudo asymmetric. C_3 in the begetter atom is called propseudo asymmetric; regardless, $H_{a'}$ and H_b are diastereotopic.

In diethyl sulfoxide. H_a and H_b are additionally diastereotopic as are Hc and HI; then again, HA is enantiotropic with H_c and H_b with H_d-Since the particle, all in all is achiral, the presence of enantiotropic iotas is conceivable.

The circumstance is completely practically equivalent to that in citrus extract in which we currently perceive H_a to be diastereotopic to H_b (as H_c is to H_d) though H_a and H_c (or H_b and H_d) are enantiotropic.

Citrus extract and diethyl sulfoxide are said to contain two enantiotropic sets of diastereotopic ligands (or two diastereotopic sets of enantiotropic ligands). In passing, we may take note of that the hydrogens connected at C_2 and C_4 in cyclobutanol additionally structure enantiotropic sets of diastereotopic hydrogens.

In the trans diol (trans-49), then again, the structure of the relating hydrogen an enantiotropic pair [C_2 versus C_4] of geminally homotopic hydrogens though in the cis 1,3-diol (cis–49) they structure a geminally diastereotopic pair of homotopic [C_2 versus C_4] hydrogens. (These certainties may become clearer when symmetry criteria are applied). The expansion measure may correspondingly be applied to perceive diastereotopic faces.

Methyl-phenethyl ketone, has a chiral focus; expansion plainly offers to ascend to diastereomers; the essences of the carbonyl carbon are diastereotopic and the C = O bunch is prochiral. This case is of significance related to Cram's standard Compounds additionally show diastereotopic faces despite the fact that the items are not chiral; have prostereogenic as opposed to prochiral faces.

The C = O group is propseudo asymmetric since C_3 is a pseudo asymmetric focus, a-Phenethyl methyl sulfide shows diastereotopic sides of an atomic plane not because of a double bond) and may, on the other hand, be viewed as an instance of diastereotopic ghost ligands (unshared matches on sulfur). This case involves chirality and the sulfur atom is prochiral.

8.10.2. Symmetry Criterion

The symmetry criteria of diastereotopic ligands or appearances are basic: such ligands or features must be connected neither by a symmetry component of the main kind (axis) nor by one of the subsequent kinds (plane, focus, substituting axis).

The students ought to persuade themselves that the even-numbered particles are either without such symmetry components or that, when such components are available, their activity doesn't serve to exchange the ligands or faces assigned as being diastereotopic.

By method for speculation, it may be called out to that in comparison to the general sort A (diastereotopic ligands-R-or faces). Type B has homotopic ligands or faces as type C. Interestingly, the ligands or faces in D and E are diastereotopic.

8.11. SPIN COUPLING NON-EQUIVALENCE

Nuclei which are an isochronous generally couple with one another as well as vary in coupling constants concerning a third core. Such non-comparability regarding turn coupling might be designated "anisogamy" and the cores are "anisogamous" just as an isochronous.

Along these lines the 13C-1H turn coupling constants of the diastereotopic methylene protons in $CH_3CH(OCH_2CH_3)_2$ are 139.6 and 141.0 Hz and the diastereotopic methyl protons in $C_6HsP(CH(CH_3)_2)_2$ have 31p-IH coupling constants of 11.0 and 14.7 Hz.

However, even isochronous cores might be anisogamous. In this way, though the two fragrant protons in 1,3-dibromo–2,5-difluoro benzene, are both isochronous and isogamous and along these lines offer ascent to a solitary reverberation, the two protons ortho to bromine in p-chloro bromobenzene are isochronous however not isogamous; the equivalent is valid for the two protons ortho to chlorine.

Assessment shows that H_a will be distinctively coupled to H_b (to which it is intrinsically heterotopic) than H_c is to $H_{b'}$; along these lines, for example, $H_{a'}$, and H_c are anisogamous. The framework is of the AA'BB' type.

In the event that in a turn arrangement of the sort A_2B_2. (or on the other hand AA'BB.') substitution of one of the B's by an alternate core Z prompts a framework A_2B_2 in which the remaining homomorphic cores An are not, at this point interconvertible by an evenness activity (C, or Sn, including o) at that point B and B' are anisogamous.

However, on the off chance that the A cores stay interconvertible by a balance activity, at that point the B cores are isogamous. By this foundation, the protons in CH_2F_2 and CH_2=C=CF_2 are isogamous since they remain enantiotropic, and henceforth isochronous, in CH_2FBr and CH_2=C=CFBr (this would be genuine regardless of whether Br was a substantially attractive core).

However, in CHF =C = CHF or cis-or trans-CHF =CHF the protons are anisogamous, for though they are connected by a C_2 center in the mixes appeared (and thus are isochronous), they are not, at this point so related in CHF=C=CHBr or cis-or transCHF=CHBr.

The equivalent is valid for the methylene protons in $BrCH_2CH_2CO_2H$ (they become diastereotopic in $BrCH_2CHC1CO_2H$ or $BrCHCICH_2CO_2H$). It is of intrigue that the protons (or fluorine cores) in CH_2F_2 ought to get anisogamous in a chiral dissolvable since the enantiotropic protons in CH_2CFBr would become diastereotopic in such a dissolvable; in any case, endeavors to exhibit such anisogamy have been productive.

8.12. CONCLUSION

PIM defines a subclass among achiral particles as a rule. By definition, PIM is showed within the sight of an outer chiral influence, which empowers separation between the heterotopic ligands or planes.

The chapter describes the stereochemistry aspects related to the topicity and PIM. Different aspects related to the topicity relationships have been evaluated based on their relation to chirality and prochirality.

In order to understand the properties and behavior of the chemical compounds, it is very essential that every aspect of these compounds should be understood and analyzed. In order to do so, topicity is one of the most essential factors of the chemical compound that has been evaluated in the chapter along with the factors associated with topicity as well.

PIM or prochirality is the property exhibited by chemical compounds in which the compounds are bonded to two similar ligands and if one of which is replaced than the atom archives chirality.

This is one of the most essential factors to be considered with chemical compounds for the analysis of the chemical compounds. In order to understand the prochiral properties exhibited by chemical compounds, compounds such as carbon, carbonyl, and imine groups have been used in the chapter to analyze the PIM behavior of the chemical compounds.

Despite the fact that this is autonomous of direction inside a chiral situation, PIM is most tangible concerning a chiral surface (Ogston's three-point connection speculation). At that point, PIM might be considered as true chirality, as its connection to a chiral surface prompts diastereomeric adducts.

Although, chirality is an aspect in which compounds do not super impose with their mirror image, but the homotopic ligands are also required to be analyzed properly in order to understand the working and properties of the chiral compounds.

The complexation of a substrate at a chiral restricting site includes stereochemical complementarity between the substrate particle and the highlights of the coupling site cavity.

It is then almost certain that there is an ideal size for every one of the three groups to be appended at the comparing pockets on a superficial level. This is on the grounds that the powerless dispersive powers, for the most part, interceding this cooperation require an ideal separation for efficient complexation to happen.

This suggests a chiral cavity would like to tie to either a chiral or a PIC atom. (The last especially applies if the gap is considered as giving a chiral surface, as in the Ogston system.) These ends have an intriguing bearing on organically dynamic atoms as a rule and the structure of medications specifically.

REFERENCES

1. Barnes, K., (n.d.). *Topicity in Stereochemistry: Relationships and Examples*. [online] Study.com. Available at: https://study.com/academy/lesson/topicity-in-stereochemistry-relationships-examples.html (accessed on 18 August 2020).
2. Chandrasekhar, S., (2016). Prostereoisomerism and biological activity: Possible implications for drug design. *Turkish Journal of Chemistry*, *40*(4), 535–545.
3. Eliel, E. L., & Wilen, S. H., (2008). Stereochemistry of Organic Compounds. John Wiley & Sons.
4. Goldberg, R. N., Tewari, Y. B., & Bhat, T. N., (2004). Thermodynamics of enzyme-catalyzed reactions—a database for quantitative biochemistry. Bioinformatics, 20(16), 2874–2877.
5. Nasipuri, D., (1994). Stereochemistry of Organic Compounds: Principles and Applications. New Age International.
6. Soderberg, T., (2019). 3.12: Prochirality. [online] Chemistry LibreTexts. Available at: https://chem.libretexts.org/Bookshelves/Organic_Chemistry/Book%3A_Organic_Chemistry_with_a_Biological_Emphasis_v2.0_(Soderberg)/03%3A_Conformations_and_Stereochemistry/3.12%3A_Prochirality (accessed on 18 August 2020).

INDEX

C

D

E

F

G

H